The Student Edition

Advanced Dynamic System Design and Analysis

- ♦ Digital Signal Processing
- ♦ Communication Systems
- ♦ Control Systems

PWS Publishing Company

I(T)P   An International Thomson Publishing Company

Boston • Albany • Bonn • Cincinnati • Detroit • London • Madrid • Melbourne • Mexico City
New York • Paris • San Francisco • Singapore • Tokyo • Toronto • Washington

PWS PUBLISHING COMPANY
20 Park Plaza, Boston, MA 02116-4324

I(T)P™
International Thomson Publishing
The trademark ITP is used under license

For more information, contact:

PWS Publishing Co.
20 Park Plaza
Boston, MA 02116

International Thomson Editores
Campos Eliseos 385, Piso 7
Col. Polanco
11560 Mexico C.F., Mexico

International Thomson Publishing Europe
Berkshire House I68-I73
High Holborn
Loncon WC1V 7AA
England

International Thomson Publishing GmbH
Konigswinterer Strasse 418
53227 Bonn, Germany

Thomas Nelson Australia
102 Dodds Street
South Melbourne, 3205
Victoria, Australia

International Thomson Publishing Asia
221 Henderson Road
#05-10 Henderson Building
Singapore 0315

Nelson Canada
1120 Birchmount Road
Scarborough, Ontario
Canada M1K 5G4

International Thomson Publishing Japan
Hirakawacho Kyyowa Building, 31
2-2-1 Hirakawacho
Chiyoda-ku, Tokyo 102
Japan

Acquisitions Editor: Bill Barter
Assistant Editor: Ken Morton
Editorial Assistant: Lai Wong
Production and Cover Design: Pamela Rockwell
Manufacturing Coordinator: Wendy Kilborn
Marketing Manager: Nathan Wilbur
Cover and Text Printer and Binder: Financial Publishing

SystemView by ELANIX® is the trademark of ELANIX, Inc.

Printed and bound in the United States of America.
95 96 97 98 99 -- 10 9 8 7 6 5 4 3 2 1

Acknowledgements

A special thank you is due Professor Mark A. Wickert of the University of Colorado at Colorado Springs. Professor Wickert is the author of the SystemView examples described in Section II of this book.

Professor Wickert first introduced his students to SystemView in February, 1994. Over time he has developed a comprehensive set of examples for use in his undergraduate and graduate level communications, signal processing and control courses. His success at building a library of exercises relevant to today's electrical engineering students is reflected in the examples provided in Section II.

Mark A. Wickert received the B.S. and M.S. degrees in electrical engineering from Michigan Technological University in 1977 and 1978, respectively. He received the Ph.D. degree in electrical engineering from the University of Missouri-Rolla in 1983.

From 1978 to 1981 he was a Design Engineer at Motorola Government Electronics Group, Scottsdale, AZ. His work at Motorola involved the design and test of very high speed digital communication electronic systems. In June 1984 he joined the Faculty of the University of Colorado at Colorado Springs where he is currently Associate Professor of Electrical Engineering. He is also a consultant to local industry in communication, digital signal processing, and microwave systems engineering. His current research interests include spread spectrum communications, mobile and wireless communications, statistical signal processing, and higher-order spectral techniques.

Table of Contents

Section I. SystemView Operation

SystemView Student Edition

Section II. Selected SystemView Examples

Section I. SystemView Operation

Chapter 1. Welcome to SystemView

Congratulations on your selection of SystemView, the premier dynamic system analysis tool for the simulation and design of today's engineering and scientific systems.   From signal processing, filter design, and communication systems to general mathematical system modeling, SystemView provides a sophisticated analysis engine embedded in the friendly and powerful Windows environment.

Complex systems may be conceived, designed, and tested within SystemView using only a mouse, your eyes, and most importantly, your mind. There are no computer codes or scripts to learn, no opaque syntax errors to battle -- virtually nothing to interfere with your concentration on the problem at hand.

Large systems created in SystemView may be simplified by defining groups of tokens as MetaSystems. A single token now represents a complete system or subsystem. Connections into and out of MetaSystems are the same as with any other SystemView token, simply click the mouse and a special window opens showing the complete MetaSystem. *The MetaSystem token is not functional in the Student Edition. For information on upgrading to the Professional Edition, refer to the information request form at the back of this book.*

SystemView automatically performs system connection checks, informs you of missing connections between system tokens, and takes you visually to the token. This feature is essential for the efficient diagnosis of your system.

Beyond system design and simulation, SystemView also provides a truly flexible Analysis window to examine system waveforms. Interactive data zoom, scroll, spectral analysis, scaling, and filtering are only a mouse-click away.

ELANIX wishes to maintain SystemView as the preferred personal computer tool for scientists, engineers, and mathematicians as they practice their art. We encourage your comments and suggestions regarding SystemView.

Chapter 2. Installing SystemView

The SystemView Setup program will automatically decompress and copy the required files to your hard drive. When the installation is complete, this program will automatically build the SystemView program group in the Windows Program Manager.

Before you run Setup, be sure that your hard disk has at least three MBytes of free space for the SystemView files.

Follow these steps to install SystemView:

> **1.**   Insert the SystemView distribution disk in the appropriate floppy drive.
>
> **2.**   In the Windows Program Manager, pull down the File menu and select Run. A dialog box will appear prompting you for the name of your application.
>
> **3.**   Type either **a:setup** if you are installing from drive A and press enter, or type **b:setup** if you are installing from drive B.

Follow the instructions which are displayed in dialog boxes that appear during the installation process.

Chapter 3. Getting Started

Now that you've installed SystemView, let's become familiar with its tools and try some examples.

3.1 Starting SystemView

During installation SystemView Setup automatically creates a SystemView program group in the Windows Program Manager. To start SystemView take one of the following actions:

❏ If the SystemView icon is selected, press Enter, or

❏ Place the mouse pointer on the SystemView icon and double click with the left mouse button.

3.2 The SystemView Environment

3.2.1 The System Window

The SystemView Design window is shown in Figure 3.1. The Menu bar contains pull-down menus from which key SystemView functions are accessed. These are summarized for quick reference in Appendix B.

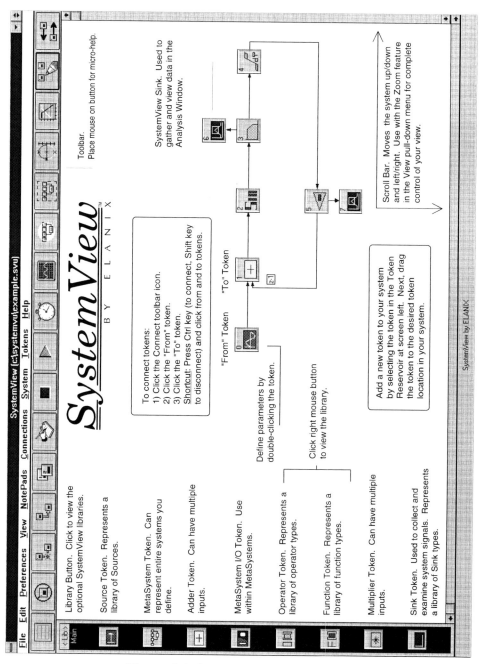

Figure 3.1 SystemView Design Window

The Action Bar consists of a group of icon buttons which act on tokens in the Design window, and two special icon buttons which are used to define System Time and to access the Analysis window. (These buttons are summarized for quick reference in Appendix A)

In SystemView the Action sequence is: **1)**Click the desired Action button (such as Connect) and then, **2)** Click on the token(s) that are to receive the action.

The **System Time button** (looks like a stopwatch) on the Action bar opens the System Time window for simulation time specification. Chapter 4 covers this button in detail.

The **Analysis button** (looks like an oscilloscope screen) is used to access the data Analysis window. After executing a simulation, click this button.

The **Message Area** at the bottom of the window displays simulation status information and token descriptions. You can always display token parameters by positioning the mouse arrow on the desired token. **Tip:** *If you click and hold the right mouse button the token parameters will also be displayed in a pop-up window.*

The Design window is the area in which the simulation design action takes place. It is here that you will launch and define the tokens and connect them for your simulation.

3.2.2 The Token Reservoir

Tokens are the building blocks of all SystemView simulations.   The Token Reservoir, on the left hand side of the Design window, contains eight generic tokens representing the eight different token classes:

Source Token:

Represents a library of Sources used to generate the inputs to your systems. This token has outputs only.

MetaSystem Token:

This token represents a (possibly very large) group of tokens which you may use as a complete subsystem, function, or process in your simulation. *The MetaSystem token is not functional in the Student Edition. For information on upgrading to the Professional Edition, refer to the information request form at the back of this book.*

Adder Token:

Forms the sum of its inputs.

MetaSystem I/O Token:

This token defines the input and output nodes for a MetaSystem. *The MetaSystem token is not functional in the Student Edition. For information on upgrading to the Professional Edition, refer to the information request form at the back of this book.*

Operator Token:

Represents a library of Operators which use the input data as the argument for the operation.

Function Token:

Represents a library of Functions which use the input data as the argument to the function.

Multiplier Token:

Forms the product of its inputs.

Sink Token:

Represents a library of Sinks for collecting, displaying, analyzing, and outputting (to an external file, if desired) the signals within a system. Sinks have inputs only.

3.2.3 Defining Tokens

You can add any of the generic token types to your simulation by double clicking the mouse on a generic token in the Token Reservoir. This will cause the generic token to be launched into the Design window. When you double click on the new token in the Design window, the token library for that token type will appear. An example window for the Function token library is shown in Figure 3.2

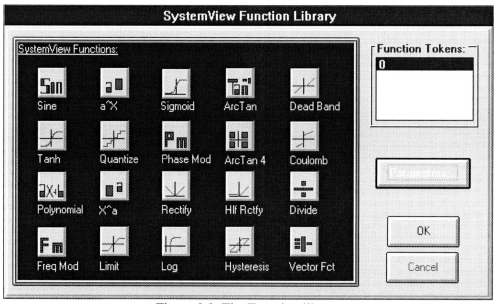

Figure 3.2 The Function library

Here you select the specific token type from the library by clicking your selection, and define its parameters by clicking the Parameters button. **Tip:** *double click the token type to bypass the Parameters button.*

A typical parameter window (in this case, for the Quantizer Function) is shown in Figure 3.3. Here you enter the specific parameters for your token. Note that these same parameters can be assigned to other tokens of the same type by selecting the desired tokens in the Apply To list.

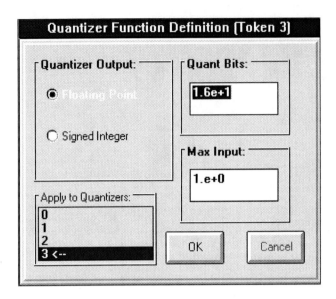

Figure 3.3 The Parameter entry window

You may perform algebraic operations while entering parameter values. For example, entering the line: (3*twopi)^2 gives 3.55058e+2 as the parameter value.

TIP: *Once a token has been defined, you can bypass the token library screen and go directly to the Parameter window by pressing the Alt key while double clicking the token.*

3.3 Your First System

Now that you are familiar with some of the basics of SystemView, lets put together the example system shown in Figure 3.4. This system will generate a sinusoid and then produce its square.

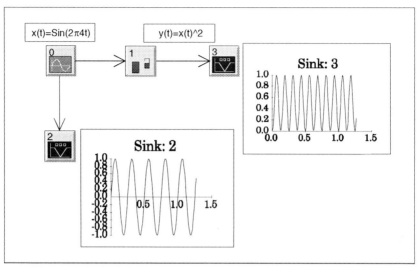

Figure 3.4 A simple system to create the square of a sinusoid.

1. Click the System Time button (the stopwatch). Set the Number of Samples to 128 (or simply click the Set For FFT button)

2. launch a Source token. Double click the token to display the Source Library. Click on 'Sinusoid'. Click the Parameters button. In the Frequency text box enter the value 4. Click OK. You have defined a unit amplitude sine wave with a 4 Hz frequency.

3. Now launch a Function token. As with the Source token, double click the token to display the Function Library. Select X^a. Click the Parameters button. In the text window type the value 2. This token will be used in our system to produce the square of the input sine wave.

4. Launch a Sink token. Double Click and select SystemView as the Sink type.

5. Connect the Source token to the Function token and to the Sink token.

6. Launch another Sink token and define it as a SystemView Sink.

7. Connect the Function token to this second Sink token.

8. Execute your system by clicking the Execute button.

9. Expand either SystemView plot by simultaneously depressing the Ctrl key and holding the mouse button down. Now drag the mouse and release to view the expanded plot.

10. Click the Analysis button (looks like an oscilloscope). Click on New Data. Click OK when the Selection Menu appears. You should see two plots, one being the input 4 Hz sine wave, and the second being the square of the first.

11. Click the FFT button. Two new plots will appear showing the spectrum associated with the two input signal. Verify that one line occurs at 4 Hz, and the other (the squared signal) occurs at 8 Hz.

3.4 Saving And Recalling Your Systems

Now that you've finished your design and examined the results, save your system by selecting Save from the File menu. You may enter any legal DOS file name. The file extension ".svu" is automatically added if no extension is typed. When working with a file that has been previously saved, you can re-save the file by selecting Save from the File pull-down menu. SystemView will automatically over-write the existing file. It is good practice to save your work from time to time

Chapter 4. System Time

System time control is accessed by clicking on the System Time button located at the top center of the system screen (it looks like a stopwatch). Figure 4.1 shows the time specification window as it will appear.

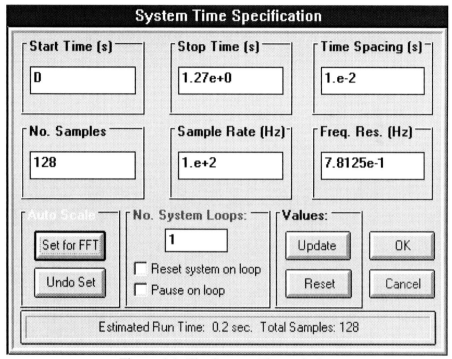

Figure 4.1 The System Time window

4.1 Start Time / Stop Time

These values control the time domain range. There are essentially no constraints on either in terms of range. The only requirement is that the Stop Time be greater than or equal to Start Time.

4.2 Sample Rate / Time Spacing

These two values control the time step used in the system simulation. Remember, SystemView is inherently a discrete-time system. You can specify either the Time Spacing or the Sample Rate. Note that these two items are *not* independent, but are constrained by the relation *Sample Rate = 1/ Time Spacing*. Changing one of these automatically forces the appropriate change in the other.

4.3 No. Samples

This entry specifies the number of time samples which will be executed by the system. The basic relation is *No. Samples = (Stop Time - Start Time) x Sample Rate+1*. As this equation relates three variables, the following rules apply; 1) If you change the number of samples, the Start Time will not change, but the Stop Time will change accordingly, 2) if you change either (or both) the Start Time and Stop Time, the number of samples will change, 3) obviously only an integer number of Samples is allowed. If the basic relation does not produce an integer, the No. Samples window will be appropriately rounded to the nearest integer. The system time will begin at Start Time and execute the designated number of samples. Note that the Sample Rate remains fixed unless you directly make the change yourself.

4.4 Frequency Resolution.

This entry is the frequency resolution obtained from a time series when a Fourier Transform is applied to the data. This value is computed from the basic formula, *Frequency Resolution = Sample Rate / No. Samples*.

4.5 Update Values

When a change is made to a particular time specification, all of the related time specifications are automatically updated when the Update button is clicked. Clicking the OK button has the same effect.

4.6 Auto Scale

This feature is to facilitate Fourier Transform operations. The SystemView FFT routine uses a radix 2 basis for optimal speed. It will pad zeros if the number of samples is not a power of 2. The Auto scale feature allows you to easily set the No Samples field to a power of 2. By clicking on this button, the No. Samples is rounded to the nearest power of 2. To do this the Stop Time is adjusted automatically. The Undo Set returns you to the original setting, if for some reason, the FFT set is not wanted.

4.7 Number of System Loops (Reset / Pause System on Loops).

This is a powerful feature of SystemView. It provides the capability for automatically repeating runs of your system.

It is best to discuss the use of the Reset System On Loop option first. This feature controls what happens to your system after each Loop. If the feature is off, then all system parameters will be remembered from Loop to Loop. If, for example, there is an integrator in your system, the output at the end of one Loop simply becomes the initial condition for the next Loop. In this way a large time simulation, as described in the first bullet below, can be performed in a time-seamless manner.

If the Reset feature is activated, then all token values are reset after each Loop. Thus statistical averaging, as described in the second item below, is possible.

❑   You have an external data file having N samples and N is greater than 32,000 (The Student Edition is limited to 32,000 samples/loop). You can circumvent the 32,000 sample restriction as follows. Set the No. Samples = M ($M <$ 32,000). Now set the No. System Loops = L, such that $L \times M = N$ ($N > 32,000$). SystemView will process the total number of samples N.

❑   You wish to make repeated runs for the purpose of averaging the results over many trials. This feature is used with the Averaging Sink. For example, suppose the basic system is a sine wave plus noise which is followed by an FFT. The SNR in the signal FFT bin can be improved by setting the No. System Loops to some value L. Each run will use a different noise segment. The Averaging Sink simply adds the individual time segments together. The signal portion adds coherently, while the noise deviation grows as the square root of L.

SystemView Student Edition

The Pause On Loop feature allows you to pause the simulation at the end of each loop and analyze the current results. For example, you can go to the Analysis window and observe the current sink waveforms.

Chapter 5. Filters and Linear Systems

The Linear System token in the Operator library is one of the most versatile and powerful of the SystemView tokens. Simply put, it provides the means for defining any linear system transfer function. However, embodied in the definition of the token are a wide range of options to define windows and filters, including several finite impulse response (FIR), and infinite impulse response (IIR) types. In addition, one can define a complete linear system in the z-domain or the s-domain (Laplace transform) with arbitrary poles and zeros.

5.1 Linear System Token Definition

Launch a Linear System token by double clicking on the generic Operator token in the Token Reservoir. Double click the new generic token to access the Operator library. A Linear System is selected by double clicking the Linear System token in the library or by clicking the library token and then clicking the Parameters button.

The Linear System Definition window is shown in Figure 5.1.

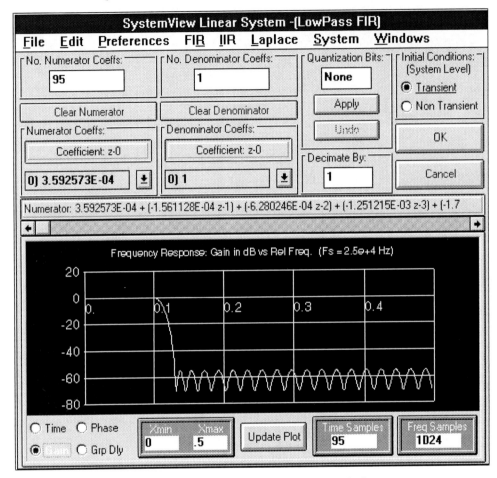

Figure 5.1 The Linear System design window.

The transfer function can be viewed in algebraic form just above the scroll bar. The scale of the plot region can be change by adjusting xMax, xMin, Time Samples or Frequency Samples. Clicking the Update button enforces these changes. The Preferences Menu allows you to eliminate the grid in the plot window. **Tip:** *Double-click in the plot area to toggle the grid on and off.*

The final result of the system specification is a z transform transfer function H(z), of the form

$$H(z) = \frac{a_0 + a_1 z^{-1} + \dots a_n z^{-n}}{b_0 + b_1 z^{-1} + \dots b_m z^{-m}}$$

Linear System tokens may be defined in several different ways. The options are:

❏   Manual entry of the z-domain coefficients $\{a_k, b_k\}$.

❏   Importing the coefficients $\{a_k, b_k\}$ from an external file.

❏   Design of a FIR filter.

❏   Selection of an IIR filter from a library.

❏   Specification of your system in the Laplace s-domain.   SystemView will automatically compute the z-domain coefficients.

5.2 Manual Entry of Coefficients

You can enter up to 1024 coefficients into both the numerator $N(z)$ and the denominator, $D(z)$ of the transfer function,

$$H(z) = \frac{N(z)}{D(z)}$$

To manually enter coefficients in the numerator, select the text boxes that define the number of coefficients for the numerator and denominator, respectively, and enter the desired number of coefficients for each. Next select the numerator coefficient text box and enter each coefficient. Each successive coefficient is added by using the down arrow key or by pressing <Enter>. The coefficients for the denominator are added in the same way.

Once you have completed entry of the coefficients, the impulse response of the transfer function can be plotted versus normalized time by selecting the Time button on the plot window, or versus normalized frequency by selecting Gain. The default plot is time. Likewise, the Phase response and the Group Delay can be plotted by clicking the appropriate button.

System coefficients may be saved by selecting the Save Coefficient File item from the File menu at the top of the Linear System design window.

As a simple example enter the following set of coefficients in the numerator: 1, 0.25 and 0.01. In the denominator enter 1 and 0.5. Remember you must first enter the number of coefficients in the No. of Numerator Coeffs text box (in this example, 3), and the No. of Denominator Coeffs text box (2). The resulting transfer function is:

$$H(z) = \frac{1 + 0.25z^{-1} + 0.01z^{-2}}{1 + 0.5z^{-1}}$$

When you complete the entry of these coefficients, click on the Update button at the bottom of the display. The impulse response will appear in the plot area and should appear as shown in Figure. 5.2. You can examine the frequency response by selecting the Gain button and it should appear as shown in Figure. 5.3.

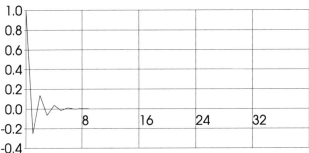

Figure 5.2 Impulse response

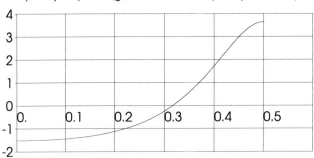

Figure 5.3 Gain response

As a second example of the custom IIR capability, consider the expression describing the Simpson's rule numerical integration:

$$I = \Delta \cdot (y_1 + 4y_2 + y_3)/3$$

which approximates to order Δ^5 the integral of a function y using three points separated by step size Δ. To extend this relation to include more than three points the relation becomes,

$$I_n = I_{n-2} + (y_n + 4y_{n-1} + y_{n-2}) \cdot \Delta / 3$$

Taking the z transform of the above relation gives the transform of the integrator as (ignoring the scale factor $\Delta/3$),

$$H(z) = \frac{1 + 4z^{-1} + z^{-2}}{1 - z^{-2}}$$

Set up a linear IIR system as described above. Normalize using a Gain token of value Time Step/3. Drive this system with any input source. Verify that the output is the integral of the input. This idea can be extended indefinitely to even higher orders of integration.

5.3 Importing System Coefficients for Custom Design

You can import the Linear System coefficients (numerator and denominator) from an external file, one in which the filter coefficients have been calculated previously by some other means. In order to properly import the coefficients the structure of the file to be imported must be as follows:

❏   The data must be in text format (ASCII) or 32 bit binary format.

❏   The data for the numerator is first. This data must have a header defining the number of coefficients in the numerator, directly followed by the coefficients.

❏   The column continues without a break by defining the number of coefficients in the denominator, followed by the denominator coefficients.

Consider, for example, the following three text data files (the "=" character is the only required identifier in the header of each list),

File 1	File 2	File 3
Numerator=3	N=3	=3
1.0	1.0	1.0
.25	.25	.25
.01	.01	.01
Denominator=2	D=2	=2
1.0	1.0	1.0
0.5	0.5	0.5

Each of these files defines the same transfer function that we defined in the previous section. Coefficient files are also saved in the same format.

To import a file, select File in the Linear System design window, and then select Open Coefficient File from the menu. A window will appear in which the files available in the selected drives are viewed. Select the desired file from this menu.

5.4 Finite Impulse Response (FIR) Filter Design

The FIR design is accessed by clicking FIR in the Menu bar. There are two groups of FIR filters. The first group consists of six classes of FIR filters:

- ❑ Low Pass
- ❑ Band Pass
- ❑ High Pass
- ❑ Hilbert Transforms
- ❑ Differentiators
- ❑ Band Reject

In each case, when you select the desired filter type, an appropriate graphical design window appears prompting you to define the pass band, transition band, and stop band of the filters. In addition, you are prompted to define the pass band ripple in the appropriate filter types.

All frequencies are specified as a fraction of the sampling rate seen by the filter. For example, if your system Sample Rate is 1 MHz and you design a 50 KHz FIR lowpass filter, the filter cutoff frequency should be entered as 0.05 (50 KHz / 1 MHz) in the FIR design window. If the filter is preceeded by a Decimator or Sampler token, then the frequencies should be a fraction of the sample rate seen by the filter.

The estimated number of taps required to realize the filter can be displayed by clicking the Update Tap Estimate button. Remember, *you must enter the desired number of taps in the No. FIR Taps text box.* This number need not agree with the estimated value. Filter coefficients are then calculated by selecting the OK button.

Upon completion of the calculation, the filter coefficients are then displayed and the filter response is plotted. The default plot is in the time domain. However you can select, gain, phase or group-delay plots as well. You can change the scale of the plot by modifying the values of xMax, xMin, and Number of Samples at the bottom of the display.

As an example let's design a lowpass FIR filter. From the FIR menu select LowPass. The LowPass Filter design window will appear, as shown in Figure 5.4. On the right of the window you will see text boxes which define the number of FIR taps, the In-Band Ripple and the maximum number of iterations to be used by the algorithm to calculate the FIR tap coefficients. (In most situations, the coefficients will converge to their proper value before the maximum number of iterations is reached.)

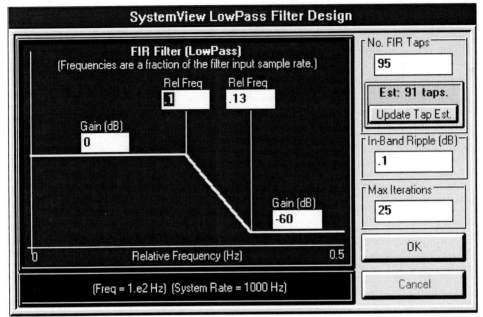

Figure 5.4 The LowPass filter design window.

In the window are text boxes with which you define the gain in the filter pass band, the corner frequencies of the pass band and transition band, and the Gain of the filter in the stop band. Let's define a low pass filter by entering the values of the filter design in the appropriate text boxes, as shown in Figure 5.4. Notice that the No. of FIR Taps text box is empty. You can enter any desired number of taps up to 1024. However, a better procedure is to click the Update Tap Estimate button. An estimate of the number of taps to fulfill the design requirements will appear in the text box directly above the button. It is wise to select this estimate as the minimum number of taps to complete the design.

When you have finished entering the design parameters, the coefficients are calculated by clicking the OK button. As coefficient computation is underway, a progress bar indicates the status of the coefficient calculation. When the coefficient computation is complete, the time domain impulse response of the filter appears in the plot area. Click the Gain option to see the plot shown in Figure 5.5.

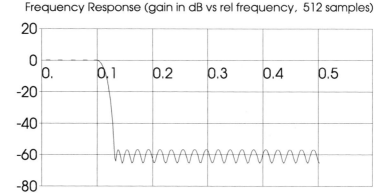

Figure 5.5 FIR Filter gain response

The other filter types are designed in the same way using the design windows.. Hilbert transforms, which introduce a 90 degree phase shift in the input waveform over a selected bandwidth, (usually that of the signal) are generated in the Hilbert Transform Design window.

The second group of FIR filters are all lowpass designs based on standard impulse responses associated with common window functions. Five types of Window FIR filters can be generated in the Linear System design window. These are:

❑ Hanning

❑ Hamming

❑ Bartlett

❑ Blackman

❑ Elanix

(The Elanix window is the auto-convolution of the Hamming window and therefore has -80dB sidelobes at the expense of a wider main lobe width.)

You begin the design by selecting the desired window filter type. Upon selection, a design window appears, showing the general shape of the filter. As with the previous FIR designs, the window contains data input areas which prompt you to enter the desired filter parameters. When the data is entered, the number of taps is estimated by clicking the Update Tap Estimate button. In this case the estimated number is the actual number used, and no other choice is available. Clicking OK generates the filter coefficients.

5.5 Infinite Impulse Response (IIR) Filter Design

You can design four classes of Infinite Impulse Response Filters using this menu selection. These are:

- ❏ Butterworth
- ❏ Bessel
- ❏ Chebechev
- ❏ Linear Phase.

These may be lowpass, highpass, or bandpass filters. When you select one of these types from the pull down menu, the appropriate design window (as shown in Figure 5.6) appears on your screen.

The general shape of the selected filter is determined by the type itself. The data required is the order of the filter (i.e., the number of poles), the 3 dB bandwidth in Hertz, and when appropriate, the pass band or phase ripple in dB.

As an example lets define a Chebyshev IIR low pass filter. Define the system Sample Rate to be 1 MHz. From the pull down menu under IIR, select Library. To define the filter, click Chebyshev, low pass, and choose the number of poles to be 9 (this type of filter can have a maximum of nine poles), an in-band ripple of 0.1 dB and a cutoff frequency of 20 KHz.

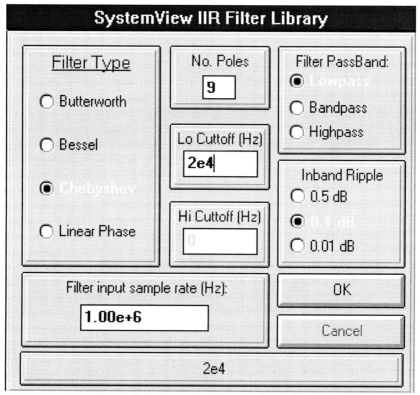

Figure 5.6 The IIR Filter design window

When you have made these entries, click OK and the impulse response of the filter will be plotted. The gain of the filter versus relative (to the system sample rate) frequency is plotted by selecting the Gain button. The gain plot should appear as shown in Figure 5.7.

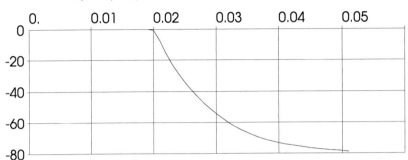

Figure 5.7 IIR Filter gain response.

5.6 Laplace Systems

SystemView provides the capability to directly implement a continuous linear system in a single token if you have its Laplace transform. The implementation architecture is to break the system into a product of fourth-order sections:

$$H(s) = \prod_{k=1}^{n} H_k(s)$$

$$H_k(s) = \frac{a_{4,k}s^4 + a_{3,k}s^3 + a_{2,k}s^2 + a_{1,k}s + a_{o,k}}{b_{4,k}s^4 + b_{3,k}s^3 + b_{2,k}s^2 + b_{1,k}s + b_{0,k}}$$

Each fourth-order section is automatically transformed into the z-domain by SystemView using the bilinear z transformation,

$$s = 2f_s\left(1 - z^{-1}\right)/\left(1 + z^{-1}\right)$$

where f_s is the sample rate seen by this token in your system. SystemView automatically senses the presence of decimators and/or sampler tokens preceding this token, and adjusts the z-domain coefficients accordingly.

The Laplace System Design window is shown in Figure 5.8. It appears when you click on the Laplace menu located at the top right of the Linear System Design window. In the example shown there is only one section having no zeros and a double pole at s = -1.

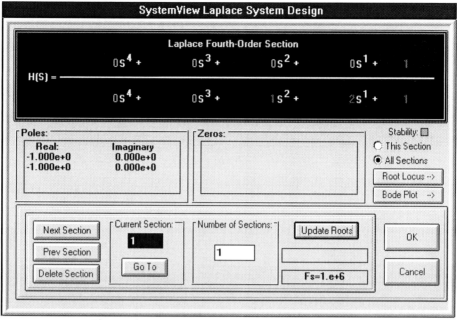

Figure 5.8 The Laplace System design window.

For example to implement a linear system having two sections,

$$H(s) = \frac{s+1}{s^2 + 2s + 2} \cdot \frac{1}{s^2 + 100}$$

1. Clear the system. Define system Time with a Sample Rate of 100, Start Time = 0, and Stop Time = 10.

2. Launch an Operator token, double click, and then double click the Linear System token in the library. Click on Laplace to produce the Laplace System design window.

3. Enter 2 in the Number of Sections window. The Section Number window will read 1.

4. Enter the coefficients for the first section in each s-coefficient text box in the graphic.

5. Click the Next button. The Section Number will change to 2.

6. Enter the coefficients for the second section.

7. Click the Update Roots button to see the system poles and zeros.

8. Click the Root Locus button to see the plot shown in Figure 5.9. Press the Ctrl key and click and drag the mouse to zoom your view of any portion of the display. Click the Exit button.

9. Click the Bode Plot button to see the plot shown in Figure 5.10. Press the Ctrl key and click and drag the mouse to zoom your view of any portion of the display. Click the Exit button.

10. Click the OK button. Note that the z-domain coefficients of this system are now in the Numerator Coeffs and Denominator Coeffs pull down windows respectively.

11. You can view the time and frequency domain characteristics of this system by clicking the buttons in the lower left corner of the window. In the time response note the oscillations have a frequency $10/2\pi$, arising from the pole at s=10.

12. Click OK in the Laplace window to return to the system Design window.

13. Finish the example by connecting an Impulse Source, and a Sink to the Linear System token.

14. Execute the system and view your results. Click the FFT button and verify the frequency location of the strong spectral line.

5.7 Root Locus and Bode Plots in the Linear System Design Window

SystemView will compute interactive Root Locus and Bode plots for your linear system. Simply click the System pull down menu and select Root Locus or Bode Plot (in the Laplace window you may use the Root Locus or Bode Plot buttons).

The Root Locus is the locus of the poles (as a function of the loop gain k) of the closed-loop feedback system whose open-loop transfer function is the linear system specified in the Linear System design window. Thus it is a plot of the poles of

$$\frac{H(s)}{1+kH(s)}$$

or, equivantly, the roots of

$$D(s)+kN(s)=0$$

ITP HIGHER EDUCATION

The enclosed materials are sent to you for your review by
OPEN TERRITORY 33

SHIP TO:
Randall Seed
Northeastern Univ
Electrical Engineeri
360 Huntington Ave
Boston MA 021150000

SHP VIA: UPS
P.O.:
DOCNUM: 08672317313
SHP ZIP: 021150173

LOCATION	QTY.	ISBN	AUTHOR / TITLE
K-ASY-033-01	1	0-534-95028-0	ELANIX SYSTEMVIEW

PKG ID: 58060G8
BAT ID: 0203323
CARTON: 1 OF 1

Date   Account Contact
Date   Account Contact

OFFL-NONFRI-DOM-SLSB-STAPL-
067/100

where

$$H(s) = \frac{N(s)}{D(s)}$$

is the linear system you have defined. Of course if your system is defined in the z domain then the above definitions are in terms of $H(z)$. An example is shown in Figure 5.9 for the fourth-order Laplace system discussed in Section 5.6.

The Bode plot is a plot of the magnitude and phase of $H(s)$ as a function of frequency f in Hz (with $s = j2\pi f$). An example is shown in Figure 5.10 for the system discussed in Section 5.6.

While in either the Root Locus or Bode Plot window you may zoom the display by pressing the Ctrl key and clicking and dragging the mouse to outline the area to be zoomed. The number of samples, the gain or frequency range, and linear or log (gain or frequency) spacing may be specified.

Please see Chapter 9, Root Locus and Bode Plots for more detailed information.

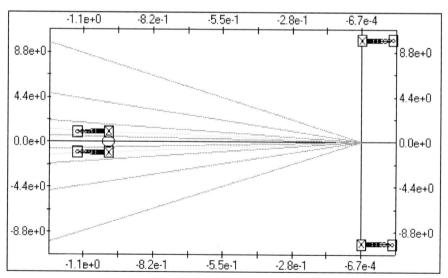

Figure 5.9 Root Locus plot for the fourth-order example.

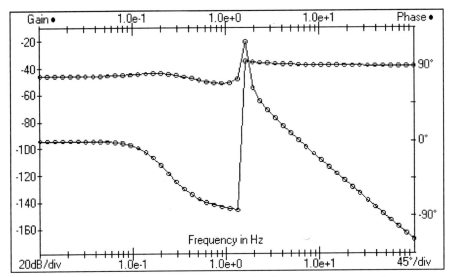

Figure 5.10 Bode plot for the fourth-order example.

5.8 Windowing

In some applications it is desirable to window existing filter coefficients. Five Window types are available: Hamming, Hanning, Bartlett, Blackman, and the Elanix window. These windows are applied after filter definition by selecting the desired window type from the Window menu. Upon selection, the window is immediately applied to the numerator coefficients. The results are plotted as the selected time, gain, phase or group delay function. The application of the window can be removed by selecting Undo Apply.

As an example let's apply the Hamming window to the lowpass filter designed in Section 5.4. To do this, select Hamming from the pull down menu under Windows. The window is immediately applied to the coefficients of the filter and the impulse response plotted. If you select Gain, you will notice that the filter ripple in the pass band has been significantly attenuated. To remove the influence of the window, select Undo Apply from the Windows pull down menu.

5.9 Coefficient Quantization

In the implementation of a digital filter in hardware, an important issue arises as to how many binary bits are required to represent the otherwise continuous filter coefficients. This is not an academic question. It is well known that the filter performance can be critically dependent on this number.

SystemView has the capability to help you make such a determination. In the Linear System design window there is a text box labeled Quantization Bits. The default is 'None'. In this case all filter design algorithms produce coefficients in full precision. To see the effects of quantizing a filter to 8 bits, for example, simply enter the text area and type in 8 and click the Apply button. The numerator and denominator coefficients are automatically quantized to 8 bits. The coefficient values in the two coefficients windows will change accordingly. The plot will also change to reflect the filter performance in the new configuration. To return to the original non-quantized condition, simply click on the Undo button.

Chapter 6. The Analysis Window

6.1 Analysis Window Procedures

You may quickly zoom any region of displayed plots by pressing the Ctrl key and dragging the mouse. The outlined region of the plots will be expanded for a more detailed view.

6.1.1 The Environment
The Analysis window is the primary vehicle for viewing your data. There are a variety of options which enhance the flexibility and usefulness of the displays. As with the rest of SystemView , these options are easily invoked through simple button clicks or pull down menus.

6.1.2 Accessing the Analysis Window and Displays
Click on the Analysis window button (looks like an oscilloscope) to access the Analysis window from the System window. Click the System button to return to the System window.

6.1.3 Translation
Their are Scroll Bars located on the right and bottom of the Analysis window. They allow you to scroll through your plot vertically and horizontally. The x and y coordinates can be controlled independently.

TIP: Press the Ctrl key and click and drag the mouse to outline a specific area of the current plot. SystemView will automatically zoom the selected region.

TIP: Click and drag the mouse anywhere within the plot to view different areas of a zoomed plot.

6.1.4 Spectral Display (FFT)
A valuable feature of the Analysis window is the ability to obtain a variety of spectral plots of your data. You only need to select the type of spectral display or FFT in the Sink Calculator, and the operation will be performed on the active plot window.

Some general guidelines for using the FFT operation in the Sink Calculator:

> The FFT is performed on 2^n points. If your time data is not a power of 2 the data will be padded with zeros to comply. You can control the number of data points by using the Set For FFT button in the System Time window. When you perform an inverse FFT on padded data, the time function will be displayed with the padded zeros.

> If you isolate a subsection of a time or spectral plot, and then select FFT, only the data currently in view will receive the operation. Thus using the Zoom feature to edit the frequency axis is equivalent to performing a filtering operation after the signal is returned to the time domain via the inverse FFT.

6.1.5 Scatter Plots

The Scatter plot feature allows you to plot two dependent time functions parametrically against each other, as opposed to time oriented line plots. A simple example will illustrate the effect.

> 1) Generate a unit amplitude sine and a cosine signal.

> 2) Enter the Analysis window and display the two functions.

> 3) In theSink Calculator, select Style and Scatter Plot. Select one of the Sinks for the X , and the other for the Y.

> 4) Click OK. The new plot window will show a circle, a plot of X vs. Y (the equation $x^2 + y^2 = 1$).

6.1.6 Plot Captions and Titles

Double click any of the plot captions to edit the caption. Under the Preferences pull-down menu, select Customize Features and Annotation. Here any of the captions may be enabled or disabled.

6.1.7 Slicing

This feature creates an overlapped (folded) plot from a single plot. A common use of this feature is to produce the so called 'eye diagrams' used in analyzing distortions in digital communication systems. When this feature is selected in the Sink Calculator the current time parameters used in the plot are the defaults. The Slice Time defaults to the total system time. Suppose the following parameters apply:

Start Time = 0 sec
Stop Time = 100 sec
Time Slice = 100 sec

The output is the normal line plot from t = 0 to t =100. By selecting Time Slice = 2 sec, the plot will be folded back on itself as follows;

Plot 1 $0 \le t \le 2$
Plot 2 $2 < t \le 4$
Plot 3 $4 < t \le 6$
      etc.

for a total of 50 plot lines. If the output was a filtered digital waveform having bit time T sec, then selecting Time Slice = T sec will generate bits folded back on each other, producing the eye pattern.

6.1.8 Copy and Paste to Other Programs

You can export your plots to other applications by using the standard Windows cut and paste techniques. The most useful application is to export your plot to a word processing or drawing application. Once there, you can annotate the drawing at will. To access this option simply select the edit pull down menu and choose Copy Active Plot Window.

6.1.9 Plot Animation

SystemView provides a powerful plot animation tool for the analysis of your dynamic systems. This tool is activated by clicking the Preferences menu in the Analysis window and selecting Customize Features.

Enabled/Disabled button: Click to enable or disable the animation feature.

Animation speed scroll bar: Used to set the speed at which the animation is run.

Show History: When this item is selected, each point plotted will remain on the screen. Otherwise, each point is erased before the next is plotted.

Manual control: When this is selected, the plot animation proceed in one of two ways. By tapping the space bar you can step through each of the points in the plot. Alternatively, when you hold the space bar down, points in the animation will continue to plot until the bar is released.

Animation is particularly informative when using the scatter plot feature with multiple sinks in a dynamic system.

6.2 Analysis Window: File Menu

6.2.1 Open Plot File
Use to open an existing plot file.

6.2.2 Save Plot Window
Saves current plot using existing name.

6.2.3 Print Plot Window
Prints the activeplot window.

6.2.4 Printer Setup
Produces Printer Setup dialog

6.2.5 Printer Fonts
Produces Font Setup Dialog.

6.3 Analysis Window: Edit Menu

6.3.1 Copy Active Plot Window to Clipboard
Copies the bitmap image of the current plot to the clipboard.

Chapter 7. Time Delays In Feedback Paths

This chapter is intended for the more advanced user who uses feedback loops as part of their system. It discusses in some detail the important timing issues that you should be aware of to ensure proper results.

Consider the simple analog feedback system shown in Figure 7.1.

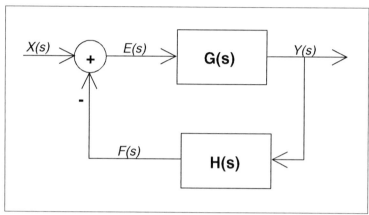

Figure 7.1 A continuous feedback system.

The basic Laplace equations describing this system are,

$$Y(s) = G(s)E(s)$$

$$E(s) = X(s) - F(s)$$

$$F(s) = H(s)Y(s)$$

resulting in the loop transfer function,

$$\frac{Y(s)}{X(s)} = \frac{G(s)}{1 + G(s)H(s)}$$

Now consider the same loop, only this time you wish to implement the equations as a sampled data system on a computer. The equations are,

$$y(t_k) = g(t_k) \otimes e(t_k)$$
$$e(t_k) = x(t_k) - f(t_k)$$
$$f(t_k) = h(t_k) \otimes y(t_k)$$

where $h(t_k)$ and $g(t_k)$ are the time impulse responses of the forward and feedback loop transfer functions, and $\otimes$ indicates convolution.

To calculate $y(t_k)$, we must calculate $e(t_k)$. To do this we use the second equation. Finally, to calculate $f(t_k)$, we go to the last equation. However, to compute $f(t_k)$ we must know $y(t_k)$, which is where we started. Thus we see that the set of equations cannot be implemented as they stand. The fundamental issue is that in a feedback loop a value which is fed-back is, in turn, dependent on its own value used elsewhere in the loop. We cannot calculate the value until we calculate the value!

Two points are in order here. The first is related to the Laplace representation used at the beginning of this discussion. The time problem does not appear here since we are working with analytical representations, not physical quantities. Secondly, a real analog loop would also have the same problem. There is always a delay in the feedback connection due to the finite time of propagation through the interconnections. This rarely causes a problem in practice since the delay is usually small with respect to the inverse of the system bandwidth.

SystemView *automatically indicates the location of the implicit delay with a small box enclosing the z^{-1} indicator.* You may edit the execution sequence to move the location of the implicit delays. See Chapter 9 for more details.

The solution to this dilemma is to use a one sample delayed value of one of the parameters. For example, replace the second equation above with,

$$e(t_k) = x(t_k) - f(t_{k-1})$$

Now there is no computational problem, but there is an operational consideration. The loop now includes a one sample delay in the feedback path. To the extent that the system sample rate is large compared to any other response time, this will not be a problem. In some instances this sample delay may produces unexpected results if not considered.

For example, consider the feedback system whose forward path and feedback paths contain *fixed constants* of value G = 2 and H = 3. Then the transfer function is a constant value independent of time,

$$\frac{G}{1+G\cdot H}=\frac{2}{7}$$

Now implement this simple loop in SystemView. Drive the loop with a constant step function of unity amplitude. Set the Time control to 100 Hz sample rate, and set the stop time to .2 sec. Execute the loop. Now Examine the output of the loop. The above Laplace equations predicts that the output should be a constant value 2/7. However as you can see, this is not the case. In fact, the loop is actually unstable. The reason is that there is a 0.01 sec delay (one sample) introduced into the feedback loop as discussed above. The actual transfer function of the loop is then,

$$\frac{Y(z)}{X(z)}=\frac{2}{1+6z^{-1}}$$

Note that this transfer function has a pole outside the unit circle (z = -6) and is therefore unstable.

To verify the above assertion, launch an Operator token on the same screen. Define it to be a Linear System. Directly implement the above transfer function by entering the above coefficients in the z-domain coefficient windows. Now drive this token with the same input. Compare the outputs of the two systems, and verify that they are identical.

There is at least one instance where the sample delay can be mitigated. If your system requires a delay in a feedback path, as with an IIR filter for example, simply implement this delay with a value of the desired value minus one sample time. Remember, if you change the system sample rate, you must adjust the delay accordingly (digital systems in SystemView may use the Sample Delay Operator token, whose delay is always exactly one sample, independent of the system sample rate).

Chapter 8. The Token Libraries

 SystemView Function Library

8.1

Token	Name	Parameters	Description
	a^X	(1) Base (default = e).	Raises the specified base to the input power according to, $$y(t) = a^{x(t)}$$
	ArcTan	(1) Output Gain	Produces the inverse tangent of the input according to, $$y(t) = G \cdot Tan^{-1}(x(t))$$ where G is the gain and, $$-\pi/2 \le y(t) \le \pi/2$$
	ArcTan 4	(1) Select output as modulo or unwraped. (2) Output Gain	Produces the four-quadrant inverse tangent of the two inputs, $$y(t) = G \cdot Tan^{-1}(x_2(t)/x_1(t))$$
	Coulomb	(1) Slope (2) Y-intercept	Produces its output according to, $$y(t) = a \cdot x(t) + b \cdot sign(x(t))$$ where a is the specified slope and b is the y-intercept.
	Dead Band	(1) Dead band limit	Produces its output according to, $$y(t) = \max(0, x(t) - z), \ x(t) \ge 0$$ $$= \min(0, x(t) + z), \ x(t) < 0$$ where z is the specified dead band limit.
	Divide	(1) Output Gain	Produces its output according to, $$y(t) = x_1(t)/x_2(t), \ x_2(t) \ne 0, \ else$$ $$= x_1(t - \Delta t)/x_2(t - \Delta t) \ x_1(t), x_2(t) = 0$$ where $x_1(t)$ and $x_2(t)$ are the specified numerator and denominator.

SystemView Function Library (Continued)

Token	Name	Parameters	Description
Fm	Freq. Mod	(1) Carrier amplitude. (2) Carrier freq. (Hz). (3) Carrier phase (deg). (4) Modulation gain.	Frequency modulates a carrier with the input signal according to, $$y(t) = A Sin \left(2\pi \left[f_c t + G \int_{t_{start}}^{t} x(\alpha)d\alpha \right] + \theta \right)$$ where A is the specified amplitude, f_c is the carrier frequency, G is the modulation gain, and θ is the carrier phase offset.
	Half Rectify	(1) Zero point.	Half-wave rectifies the input according to, $$y(t) = x(t) - z, \ \ x(t) \geq z$$ $$y(t) = 0 \ \ elsewhere$$ where z is the specified zero point.
	Hysteresis	(1) Bandwidth. (2) Backlash (3) Slope.	Provides a hysteresis transfer function whose characteristics are specified by the bandwidth and gain parameters. Very small bandwidths (relative to the system sample rate) result in a 'sluggish' hysteresis (i.e., a lowpass filter), whereas large bandwidths (e.g., ~0.3 x sample rate) result in a 'stiff' hysteresis.
	Limit	(1) Input max. ($\pm$). (2) Output max. ($\pm$).	The output is limited to $\pm$ the Output Maximum when the input exceeds $\pm$ the Input Maximum. $$y(t) = (Out\,Max \,/\, In\,Max)x(t), \ |x(t)| \leq In\,Max$$ $$= Out\,Max \cdot sign(x(t)), \ elsewhere$$
	Log	(1) Log base (default is e).	Produces the logarithm of the input using the specified base. $$y(t) = Log_b \big(x(t) \big)$$

SystemView Function Library (Continued)

Token	Name	Parameters	Description
P m	Phase Mod	(1) Amplitude. (2) Carrier frequency. (3) Carrier phase. (4) Modulation gain.	Phase modulates a carrier with the input signal according to, $$y(t) = A Sin\left(2\pi\left[f_c t + Gx(t)\right] + \theta\right)$$ where A is the specified amplitude, f_c is the carrier frequency, G is the modulation gain, and θ is the carrier phase offset.
⬛X+⬛	Polynomial	(1) Polynomial coefficients (fifth-order, six coefficients)	Creates a fifth-order polynomial function of the input using the specified a_i, $$y(t) = a_5 x^5(t) + a_4 x^4(t) + a_3 x^3(t) +$$ $$a_2 x^2(t) + a_1 x(t) + a_0$$
	Quantizer	(1) Number of binary bits. (2) Input max. ($\pm$).	Quantizes the input amplitude to the specified number of levels, $$levels = 2^{bits},$$ saturating at $\pm$ the specified input maximum.
	Full Rectify	(1) Zero point.	Rectifies the input according to, $$y(t) = \left|x(t) - z\right|$$ where z is the specified zero point.
	Sigmoid	(1) Shape factor.	The sigmoid transfer function is defined as, $$y(t) = \frac{1}{1 + \exp(-2\beta x(t))}.$$ where β is the specified shape factor.

SystemView Function Library (Continued)

Token	Name	Parameters	Description
	Sine	(1) Phase offset (deg).	The sinusoidal function is defined as, $$y(t) = Sin(x(t) + \theta)$$ where θ is the specified phase offset.
	Tanh	(1) Shape factor.	The hyperbolic tangent transfer function is defined as, $$y(t) = \frac{1 - \exp(-2\beta x(t))}{1 + \exp(-2\beta x(t))}$$ where β is the specified shape factor.
	Vector Fct.	(1) Select output as the Average, Order Statistics, Modulus, or Geometric Mean. (2) Output gain or Percentile.	Produces the output according to: $$y(t) = \frac{G}{N} \sum_{i=0}^{N-1} x_i(t), \; Average$$ $$= x_i(t), the\,input\,having\,the\,specified\,rank.\; Ordr\,Stat$$ $$= G \sqrt{\sum_{i=0}^{N-1} x_i^2(t)}, \; Modulus$$ $$= G \left(\prod_{i=0}^{N-1} x_i(t) \right)^{\frac{1}{N}}, \; Geometric\,Mean$$
	X^a	(1) Exponent value.	Raises the input to the specified power, $$y(t) = x^a(t)$$

8.2 **SystemView Operator Library**

Token	Name	Parameters	Description
	Decimator	(1) Decimation factor (integer)	Down-samples the input signal by the specified factor, $$y_n = x_n, \quad n \bmod(N) = 0$$ where N is the decimation factor.
	Delay	(1) Time delay (sec).	Delays its input according to, $$y(t) = x(t - \tau)$$ where τ is the specified delay.
	Derivative	(1) Gain.	Forms the derivative of the input, $$y(t) = \frac{d}{dt} x(t), \; t > t_{start}$$ $$= 0, \qquad t = t_{start}$$
	FFT	(1) Select output as Real, Imaginary, Magnitude, or Phase of the FFT. (2) FFT size (samples)	Performs sequential FFTs on the input. Output begins after the first time window, and is continuous thereafter. (Spectral analysis of system data is best done in the SystemView analysis window by simply clicking the FFT button.)
	Gain	(1) Gain value.	Multiplies the input by the specified value. $$y(t) = G \cdot x(t)$$
	Hold	(1) Select hold last value or hold zero between samples. (2) Gain	Used to return signals to the system sample rate after being sampled or decimated.

SystemView Operator Library (Continued)

Token	Name	Parameters	Description
	Integrator	(1) Integration Order (zero or first order). (2) Initial condition (default is zero).	Forms the integral of the input, $$y(t) = \int_{t_{start}}^{t} x(\alpha)\,d\alpha + I_0,\ \ t > t_{start}$$ $$= I_0,\ \ t = t_{start}$$ where I_0 is the specified initial condition.
	Linear Sys	Several parameters, depending on your selection of Linear System type.	FIR, IIR, and Laplace system design and specification. One of the most powerful SystemView tokens.
	Digital Scaler	(1) Input word size (bits). (2) Retained significant bits.	Extracts the specified number of significant bits from the input word (sample). The input should be integer valued. For example, with integer 13, (1101)b input and 2 as the specified retained bits, the output is 3, (11)b
	Negate	None.	Forms the negative of the input. $$y(t) = -x(t)$$
	Modulo	(1) Modulus base.	Performs the modulo operation, $$y(t) = x(t) - base \cdot Int\{x(t)/base\}, x(t) \geq 0$$ $$y(t) = base + x(t) - base \cdot Int\{x(t)/base\}, x(t) < 0$$
	OSF	(1) Time window (sec). (2) Output rank (%)	Performs order statistics filtering (OSF). The output is that input sample having the specified rank within the current time window, (e.g., Rank = 50% is the median filter, Rank = 100% outputs the maximum value in the window)

SystemView Operator Library (Continued)

Token	Name	Parameters	Description
	PID	(1) Proportional gain. (2) Integral gain. (3) Derivative gain	Produces its output according to: $$y(t) = G_P x(t) + G_I \int_0^t x(\alpha)d\alpha + G_D \frac{d}{dt} x(t)$$
	Sampler	(1) Sample Rate (Hz). (2) Sample Aperture duration (sec). (3) Aperture Jitter (sec)	Performs a true sampling operation at the specified rate. The output sample is a linear combination of the input samples falling within the sample time aperture. The aperture start time is uniformly distributed over the aperture jitter time.
	Data Switch	(1) Threshold	The output is switched between the two inputs according to the control input $c(t)$, $$y(t) = x_1(t), \ c(t) < Thr$$ $$= x_2(t), \ c(t) \geq Thr$$
	Logical AND	(1) Threshold (2) TRUE output (3) FALSE output	Performs the logical AND operation on the two inputs, $$y(t) = T, \ x_1(t) \geq Thr \ AND \ x_2(t) \geq Thr$$ $$y(t) = F, \ Otherwise$$
	Logical OR	(1) Threshold (2) TRUE output (3) FALSE output	Performs the logical OR operation on the two inputs, $$y(t) = T, \ x_1(t) \geq Thr \ OR \ x_2(t) \geq Thr$$ $$y(t) = F, \ Otherwise$$

SystemView Operator Library (Continued)

Token	Name	Parameters	Description
	Logical XOR	(1) Threshold (2) TRUE output (3) FALSE output	Performs the logical XOR (Exclusive Or) operation on the two inputs, $y(t) = T,\ x_1(t) \geq Thr\ AND\ x_2(t) < Thr$ $y(t) = T,\ x_1(t) < Thr\ AND\ x_2(t) \geq Thr$ $y(t) = F,\ Otherwise$
	Logical NAND	(1) Threshold (2) TRUE output (3) FALSE output	Performs the logical NAND operation on the two inputs, $y(t) = T,\ x_1(t) < Thr\ OR\ x_2(t) < Thr$ $y(t) = F,\ Otherwise$
	Logical NOT	(1) Threshold (2) TRUE output (3) FALSE output	Performs the logical NOT operation on the input, $y(t) = T,\ x(t) < Thr$ $y(t) = F,\ Otherwise$
Z^{-1}	Sample Delay	(1) Delay (Samples)	Delays the input by the specified number of *samples*. Unlike the Delay Operator, this Operator has a delay *time* that is dependent on the system sample rate. $y_n = x_{n-k}$ where k is the specified sample delay.

8.3 **SystemView Source Library**

Token	Name	Parameters	Description
	External File	(1) File name. (2) Data format (8-bit Integer, Binary or ASCII text)	An external source allows SystemView to process any data you desire. The file format may be either 8-bit integer (0-255), 32-bit binary, or ASCII (i.e., standard text format).
	Freq. Sweep	(1) Start freq.(Hz). (2) Stop freq. (Hz). (3) Period (sec).	Generates a swept frequency sinusoid (chirp) according to, $$y(t) = \sin(2\pi f_{start}t + \pi R(t \bmod(T))^2)$$ $$R = \frac{f_{stop} - f_{start}}{T}$$
	Gauss Noise	(1) Standard deviation or power spectral density in W/Hz. (2) Mean.	Generates a random signal having a Gaussian (Normal) distribution.
	Impulse Fct.	(1) Gain. (2) Start time. (3) Bias.	Generates a single, _unit area_, (1/dt by dt) pulse $\delta(t)$ at the specified time $$y(t) = G \cdot \delta(t - t_{start}) + bias$$ If the bias is non-zero and the start time is offset from zero, then both the step and impulse response of a system may be observed.
	PN Sequence	(1) Amplitude. (2) Symbol rate (Hz). (3) Number of levels.	Generates a pseudo-random (PN) sequence of multi-level pulses at the specified rate.

SystemView Student Edition

SystemView Source Library (Continued)

Token	Name	Parameters	Description
	PSK Carrier	(1) Amplitude. (2) Frequency (Hz). (3) Carrier phase (deg) (4) Symbol rate. (5) Number of symbols.	Generates an m-*ary* phase modulated carrier, $$y(t) = Sin(2\pi f_c t + \phi_T(t) + \theta)$$ where $\phi_T(t)$ is a PN sequence of m-*ary* phase values (between 0 and 2π), T is the specified symbol period (1/symbol rate), and θ is the carrier phase.
	Pulse Train	(1) Amplitude. (2) Frequency (Hz). (3) Pulse width (sec). (4) Bias.	Generates a periodic sequence of pulses having the specified amplitude. The pulse width determines the length of the leading pulse of each period. $$y(t) = \pm A \cdot P_T(t) + bias$$
	Sawtooth	(1) Amplitude. (2) Frequency (Hz). (3) Bias (4) Phase (deg)	Generates a periodic sawtooth waveform.
	Sinusoid	(1) Amplitude. (2) Frequency (Hz). (3) Phase (deg)	Generates a sinusoid as, $$y(t) = A \cdot Sin(2\pi f_c t + \theta)$$
	Step Function	(1) Amplitude. (2) Start time (sec). (3) Bias.	Generates a step function. Note that a single pulse or an impulse may be created by setting the bias equal to the negative of the amplitude.

54

 SystemView Source Library (Continued)

Token	Name	Parameters	Description
	Time	(1) Gain. (2) Bias.	Generates a scaled replica of the system time, $$y(t) = Gt + b$$ where t is the defined system time.
	Uniform Noise	(1) Minimum value. (2) Maximum value.	Generates noise whose amplitude values are uniformly distributed between the specified maximum and minimum.

SystemView Student Edition

SystemView Sink Library

8.4

Token	Name	Description
	Analysis	The basic SystemView sink. No signals are displayed on the System screen. *To view and analyze the sink data, click the Analysis button* (located at the top of the System screen). The Analysis window opens for detailed signal analysis.
	Averaging	When used with the System Loop time parameter, the Averaging sink computes the average waveform over all system loops. Output may be observed in the Analysis window.
	Current Value	Displays the current system time and Sink input value in real-time as the system executes. Output may also be observed in the Analysis window.
	Data List	Generates a list of the data received by the sink and displays the list on the System screen. Press the Ctrl key and drag the mouse to enlarge the data list display. Output may also be observed in the Analysis window.
	External File	SystemView output is sent to a disk file. You specify the file name and format (8 bit integer, binary, or ASCII). Recursive system I/O is possible by assigning the same file to an External Source token. Output may also be observed in the Analysis window.
	Real-Time	Displays a plot of the Sink input value in real-time as the system executes. Output may also be observed in the Analysis window.

SystemView Sink Library (Continued)

Token	Name	Description
	Statistics	Computes and displays the mean, variance, standard deviation, adjacent sample correlation coefficient, maximum, and minimum of the input data. Press the Ctrl key and drag the mouse to enlarge the list. Output may also be observed in the Analysis window.
	Stop	Stops the system simulation when the input is greater than or equal to the specified threshold value. Output may be observed in the Analysis window.
	SystemView	Displays a plot in the System window. Press the Ctrl key and drag the mouse to enlarge the plot display. Output may also be observed in the Analysis window.

SystemView Student Edition

Chapter 9. Root Locus and Bode Plots

SystemView will compute interactive Root Locus and Bode plots for your linear system as it appears in the system Design window. Simply click the Root Locus or Bode Plot button. Note that this feature is also available in the Linear System window (Chapter 5), and in the Laplace window (Chapter 5).

9.1 Root Locus

After clicking the Root Locus button in the system Design window, you next specify the domain in which SystemView computes and displays the Root Locus by clicking either S-Domain, Z-Domain, or Z-Domain with added poles at $z=0$ (resulting from the implicit sample delays as described in Chapter 7), as shown in Figure 9.1.

Regardless of the type of system in the Design window, SystemView will automatically perform the conversions necessary to compute the specified Root Locus. For example, the bilinear transform is applied to all continuous Linear System tokens if you select either Z-Domain option.

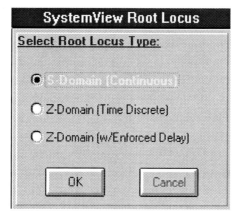

Figure 9.1 Domain selection menu for the Root Locus computations.

The Root Locus is defined as the locus of the poles (as a function of the loop gain k) of the closed-loop feedback system whose open-loop transfer function is defined in the System Design window. Thus it is a plot of the poles of

$$\frac{H(s)}{1+kH(s)}$$

or, equivalently, the roots of

$$D(s) + kN(s) = 0$$

where

$$H(s) = \frac{N(s)}{D(s)}$$

is the transfer function SystemView computes from the linear system block diagram you have designed. Of course if your system is defined entirely in the z domain then the above definitions are in terms of $H(z)$.

An example is shown for the third-order system shown in Figure 9.2. The open loop transfer function for this system is

$$H(s) = \frac{-5}{s(s+1)(s+5)}$$

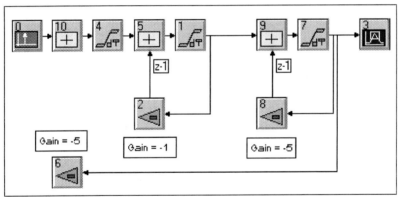

Figure 9.2 A third-order system for Root Locus and Bode computation.

The corresponding Root Locus window is shown in Figure 9.3. Note that this system will be unstable for a loop gain exceeding 30 (six times the gain value of Gain token No. 6)

The radial lines represent critical damping ratios of 0.9, 0.707 (brighter white), 0.5, 0.25, and 0.1.

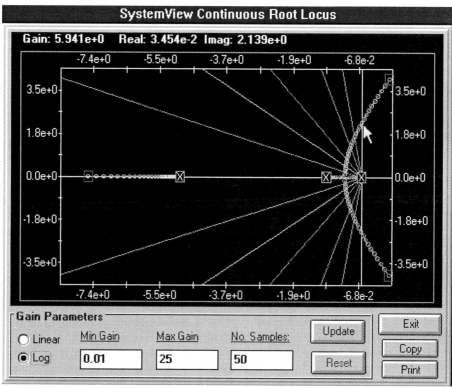

Figure 9.3 The Root Locus window for the system in Figure 9.2.

System poles are indicated in the display with the symbol 'X', and zeros with 'O'. The roots associated with the minimum gain sample (i.e., the start of the locus) are indicated with a green box, and the roots associated with the maximum gain sample are indicated with a red box.

The Root Locus window is interactive. The following is a description of the tools available to you for analysis:

❏ **Zoom**. Press the Ctrl key and click and drag the mouse to zoom a selected area.

❏ **Gain, Real and Imaginary Values**. Place the mouse anywhere on the Root Locus curve and the corresponding loop gain is displayed at the top of the window, along with the real and imaginary coordinates of the mouse.

❑ **Linear/Log**. Select either a linear distribution of the gain samples (from Min Gain to Max Gain), or a logarithmic distribution.

❑ **Min Gain, Max Gain**. Enter the minimum and or maximum gain used in the Root Locus computation.

❑ **No. Samples**. Enter the desired number of gain samples to be used in the computations.

❑ **Update**. Forces re-computation of the Root Locus. Use after changes are made in the parameter entry boxes.

❑ **Reset**. Return to the default settings.

❑ **Exit**. Exit the Root Locus window and return to the system.

❑ **Copy**. Copy the Root Locus plot to the Windows clipboard. If you have selected Use Color Printer in the Preferences menu in the system Design window, the copy will be in color.

❑ **Print**. Print the Root Locus plot. If you have selected Use Color Printer in the Preferences menu in the system Design window, the print will be in color.

When the Z-Domain option is selected in Figure 9.1, SystemView uses the bilinear transformation to convert the continuous system to the z-domain. The resulting z-domain Root Locus is shown in Figure 9.3 (the system Sample Rate is 5 Hz). Figure 9.4 shows the Root Locus when the implicit delays are included. This selection gives the most accurate stability analysis for the system simulation because it not only includes the implicit delays, but also because it is computed using the actual z-domain transfer functions used in the simulation itself (remember that SystemView is a time-discrete simulator).

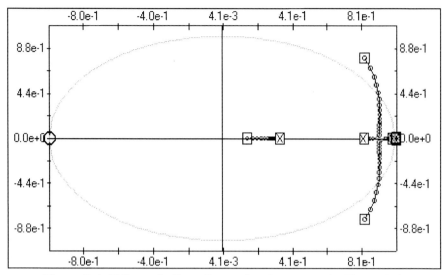

Figure 9.4 Z-Domain Root Locus for the system in Figure 9.2.

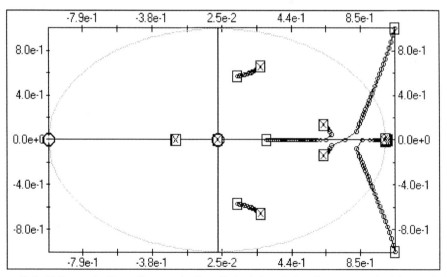

Figure 9.5 Z-Domain Root Locus (w/implicit delays) for the system in Figure 9.2.

9.2 Bode Plots

The Bode plot is a plot of the magnitude and phase of *H(s)* as a function of the frequency *f* in Hz (with $s = j2\pi f$). An example is shown in Figure 9.6 for the system discussed in Section 9.1 and shown in Figure 9.2. The Max Frequency value has been set to 100 Hz.

While in either the Root Locus or Bode Plot window you may zoom the display. Press the Ctrl key and click and drag the mouse to outline the area to be zoomed. The number of samples, the gain or frequency range, and linear or log (gain or frequency) spacing may be specified.

The system phase margin is computed from the system phase at the frequency where the system gain is 0dB. This point is automatically indicated in the display, as shown in Figure 9.6

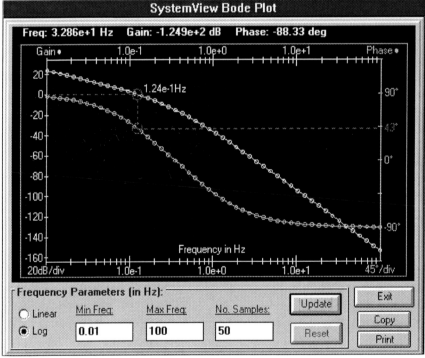

Figure 9.6 The SystemView Bode Plot Window.

Section II. Selected SystemView Examples

Professor Mark A. Wickert
University of Colorado
at Colorado Springs

Section II. Selected SystemView Examples by Professor Mark Wickert

The examples presented in this section were developed by Professor Mark A. Wickert of the University of Colorado at Colorado Springs. These examples represent actual exercises used in Professor Wickert's undergraduate and graduate level communications, signal processing and control courses. A biographical sketch of Professor Wickert is found in the Acknowledgments section of this text.

Each example begins with a problem statement and a brief development of the relevant system theory. With analytical foundations in place each example then moves into construction of a SystemView simulation and presentation of simulation results. Where appropriate, simulation results are compared to theory.

The following SystemView examples are described in this section:

Chapter 10. Signal Processing with Memoryless Nonlinearities

SystemView File: `com_nlin.svu`

Problem Statement

In this example the frequency-domain properties of memoryless nonlinearities are explored. Memoryless nonlinearities serve useful purposes in both communications and signal processing. In the study of linear systems, a memoryless nonlinearity is an example of a system that produces *nonlinear distortion*. In this context, the nonlinear distortion terms are generally considered to be undesirable.

The nonlinearity of interest here has an input/output relationship of the form

$$y(t) = a_0 + a_1 x(t) + a_2 x^2(t) + a_3 x^3(t) + \cdots \qquad (1)$$

If a_0 and a_k, $k > 1$ are all identically zero, then the system is linear and may be considered an amplifier or attenuator. Here we will consider the special case of

$$y(t) = a_1 x(t) + a_2 x^2(t) \qquad (2)$$

Only the dc bias term $a_0 = 0$ and a square law distortion term are present. In the frequency domain the output spectrum can be found using the multiplication theorem for Fourier transforms

$$Y(f) = a_1 X(f) + a_2 X(f) * X(f) \qquad (3)$$

where $X(f)$ is the Fourier transform of $x(t)$.[1] A classical textbook example (see Ziemer) is to suppose that $x(t)$ has a bandlimited spectrum of the form

---

[1] R. Ziemer and W. Tranter, *Principles of Communications*, Fourth Edition, Houghton Mifflin, Boston, MA, 1995, p. 79.

$$X(f) = A\Pi\left(\frac{f}{2W}\right) = \begin{cases} A, & f \leq |W| \\ 0, & \text{otherwise} \end{cases} \tag{4}$$

The output distortion term is of the form

$$a_2 X(f) * X(f) = 2a_2 WA^2 \Lambda\left(\frac{f}{2W}\right) \tag{5}$$

where Λ denotes the triangle function defined by

$$\Lambda\left(\frac{f}{B}\right) = \begin{cases} 1 - |f|/B, & |f| \leq B \\ 0, & \text{otherwise} \end{cases} \tag{6}$$

The composite spectrum at the memoryless nonlineartity output is thus

$$Y(f) = a_1 A\Pi\left(\frac{f}{2W}\right) + 2a_2 WA^2 \Lambda\left(\frac{f}{2W}\right) \tag{7}$$

The theoretical two-sided spectrum is sketched below in Figure 1 for the case $A = 1$, $W = 10$ Hz, $a_1 = 1$, and $a_2 = 0.1$.

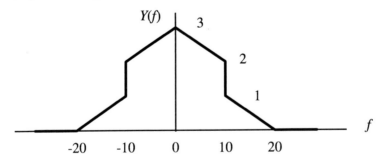

Figure 1: Two-sided spectrum at nonlinearity output.

SystemView Simulation

The results of (7) will now be simulated using SystemView. The most challenging part of the simulation is approximating the signal with bandlimited spectra. In the linear system token, under the FIR Filter Design menu, there is an option for designing truncated sinc function (sin(x)/x) filters. This is precisely what is used here. The sampling rate is set to 100 Hz and the filter cutoff frequency is approximately 10 Hz. The SystemView block diagram is shown in Figure 2.

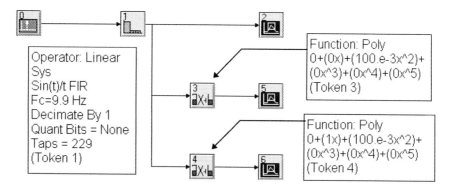

Figure 2: SystemView simulation block diagram.

From Figure 2 we see that the truncated sinc function signal is obtained by passing an impulse through the filter. The resulting signal is, of course, the filter impulse response, which is plotted in Figure 3.

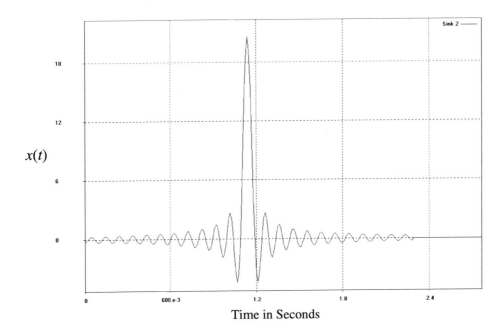

Figure 3: Truncated sinc function filter impulse response/bandlimited signal.

- Note that for realizability considerations SystemView delays the sinc function by half the total impulse response length. Here that length is 229 samples or, in time, 2.29 s.

The spectrum of the truncated sinc signal contains considerable ripple in both the passband and the stopband. In typical FIR digital filter design a truncated sinc function is windowed or shaped to reduce the ripple or ringing present at the transition band. For this application the ripple is acceptable. The Fourier transform magnitude (single-sided spectrum) of the truncated sinc signal is shown in Figure 4.

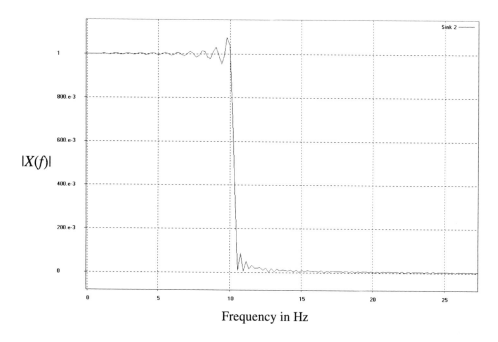

$|X(f)|$

Frequency in Hz

Figure 4: Spectrum of truncated sinc signal.

- Note that the ripple would be more evident if a dB-axis were used for this plot, but to verify (7) the linear axis is more appropriate.

The memoryless nonlinearity is implemented using a SystemView polynomial token. This token allows the user to set coefficients from a_0 up to a_5, i.e., constant up to fifth-degree coefficients. Polynomial token 3 has the nonzero coefficient $a_2 = 0.1$, thus the output signal represents the distortion spectrum of (5). The Fourier transform magnitude of this signal is shown in Figure 5.

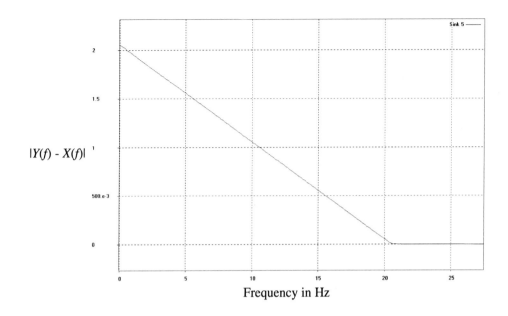

Figure 5: Distortion signal spectrum.

- Note that as predicted by the analysis of (6) the spectrum has a triangular shape and the peak is close to two.

Combining the linear term with unity gain and the square-law term with gain 0.1 is what is collected in sink token 6. The amplitude spectrum is shown in Figure 6.

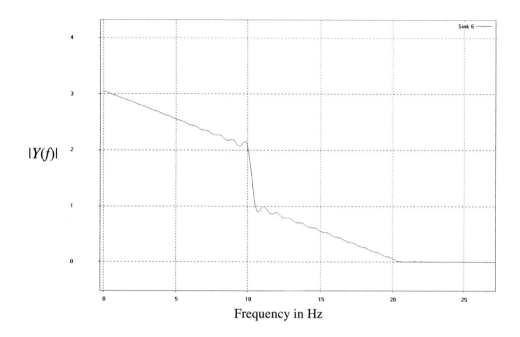

Figure 6: Composite nonlinearity output spectrum.

- The final simulation result is very close to that predicted by (7).

As a matter of practice, a square-law nonlinearity may be useful as a mixer in a communications receiver or as a frequency doubler in the generation of a stable reference oscillator.

Chapter 11. AM Superheterodyne Receiver

SystemView File: `com_amr.svu`

Problem Statement

Superheterodyne receiver techniques find wide application in radio communication systems. A basic superheterodyne receiver block diagram is shown in Figure 1.

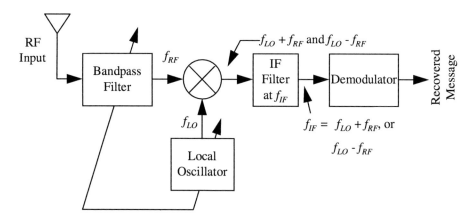

Figure 1: Basic superheterodyne receiver block diagram.

For this example the modulation scheme is assumed to be amplitude modulation (AM) on carrier frequencies of 30, 40, and 50 kHz. The intermediate frequency (IF) is chosen to be 20 kHz. The analog message bandwidth is 5 kHz or less. Note that commercial AM broadcast covers 540 to 1700 kHz and the IF frequency is typically 455 kHz. In commercial AM receivers the local oscillator (LO) is typically set above the desired RF signal, so-called *high-side* tuning. The bandpass filter at the RF input is used to reject unwanted signals and noise, the most important of which is the potential *image* signal that lies $2f_{IF}$ away from the desired f_{RF} signal. The selectivity of the receiver is accomplished with the fixed, tuned IF filter. By designing this filter to have steep skirts, energy from adjacent channels entering the demodulator can be kept to a minimum.

The frequency plan for the example system is shown in Figure 2.

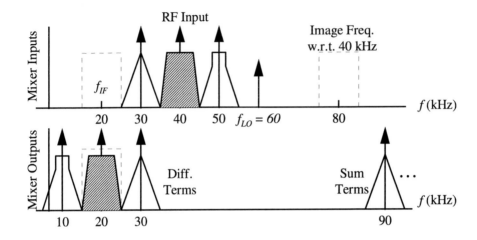

Figure 2: Frequency plan for the example receiving system.

Suppose we wish to receive the signal with the 40-kHz carrier frequency. Assuming high-side tuning, the LO is at $40 + 20 = 60$ kHz and the image frequency is at $40 + 2(20) = 80$ kHz. Following the mixing or down-conversion operation, the stations at 30 and 50 kHz are still adjacent to the desired 40-kHz signal, except now they are located at 10 and 30 kHz, respectively. The IF filter must reject energy from the adjacent channel signals yet minimize distortion on the desired signal.

For this example, the AM modulation will be recovered with a simple envelope-detection system. The total modulation depth must be at most 100%. The message signal is a swept sinusoid from dc to, at most, 5 kHz. Since the sweep time is finite, the actual message bandwidth will be greater than the stop frequency of the swept source.

SystemView Simulation

The simulation block diagram is shown below in Figure 3.

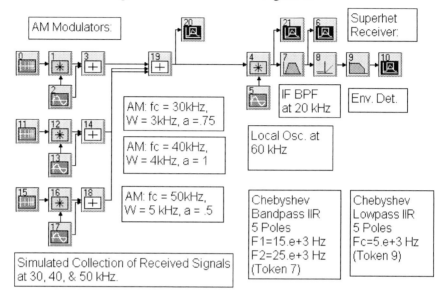

Figure 3: SystemView simulation block diagram.

The system sampling rate is 200 kHz. The tokens on the left side of the block diagram are responsible from generating the three AM signals at carrier frequencies of 30, 40, and 50 kHz. The modulation depth is controlled by summing pure carrier back in with the product of the swept sinusoid and the carrier sinusoid. To more clearly identify the transmitted signals within the receiver, each signal uses different combinations of sweep bandwidth and modulation depth. To simplify the receiver, the RF bandpass filter is not included; however, one may be added very easily. The IF bandpass filter with frequency response shown in Figure 4 is a 5-pole Chebyshev with a 10-kHz 3-dB bandwidth.

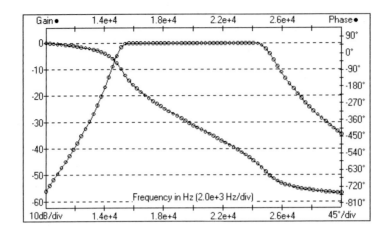

Figure 4: 20 kHz IF bandpass filter frequency response in dB.

The envelope-detection lowpass filter is a 5-pole Chebyshev with a 5-kHz 3-dB frequency.

A spectrum plot of the received RF signals is shown in Figure 5.

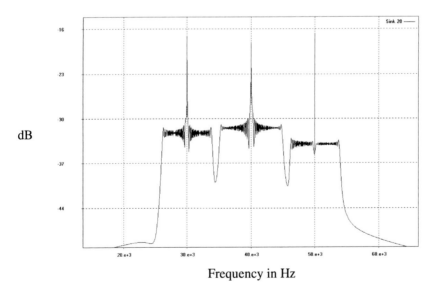

Frequency in Hz

Figure 5: Received RF signal spectrum.

The center signal, with the largest modulation depth and widest message bandwidth, located at 40 kHz is the desired signal. After mixing with a 60-kHz local oscillator signal, the spectrum shown in Figure 6 is obtained.

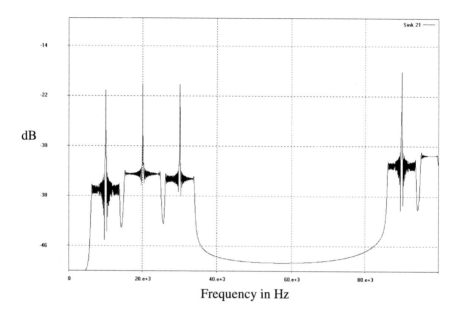

Figure 6: Mixer output spectrum when tuned to receive the 40-kHz signal.

Note that in Figure 6 we see a portion of the mixer output sum signals at 90 and 100 kHz. It can also be observed that the sum signal spectra appear to have a higher spectral amplitude than the difference spectra. The reason for this is that the *folding frequency* in the simulation is at 100 kHz; hence, the mirror image spectral components sitting at 110 and 100 kHz have produced aliasing. The spectrum of Figure 6 passes through the IF filter and produces the spectrum shown in Figure 7.

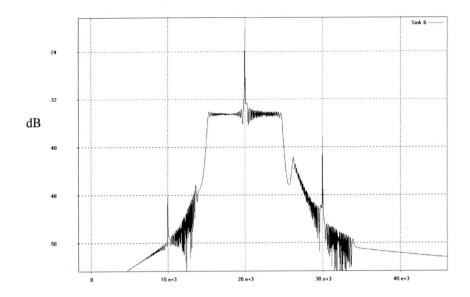

Frequency in Hz

Figure 7: IF filter output spectrum with adjacent channel signals present.

In Figure 7, note that portions of both adjacent channels are leaking through the IF filter skirts. The envelope-detector output will be dominated by the desired message signal since the carrier at 20 kHz is more than 15 dB above the 10- and 30-kHz carriers. The envelope detector time-domain output is shown in Figure 8.

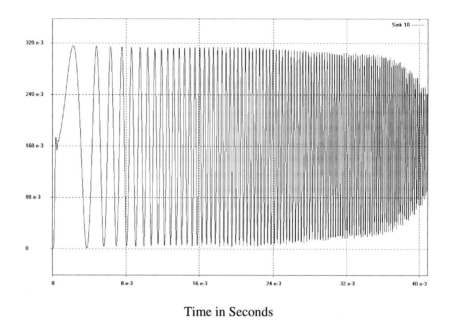

Time in Seconds

Figure 8: Envelope detector recovery of the 0- to 5-kHz swept-sinusoid message.

The amplitude rolloff as the message frequency reaches 5 kHz is due to bandlimiting imposed by both the IF filter and lowpass detection filter.

Receiver Selectivity

As a measure of the receiver selectivity we will use SystemView to measure the ratio of desired signal power to undesired signal power passing through the IF filter. This will require two special simulation runs. For the first, the 30- and 50-kHz AM transmitters will be turned off (carrier amplitudes set to zero). Using the statistics button in the Analysis Screen, the power at the IF filter output is recorded. Note for this measurement the displayed waveform should be in the time domain as shown in Figure 9.

SystemView Student Edition

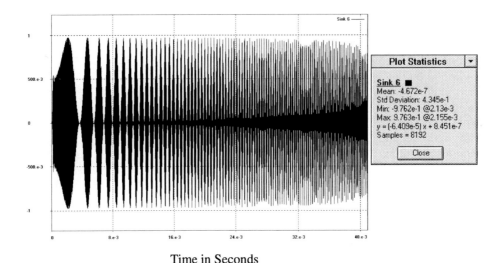

Time in Seconds

Figure 9: (a) IF filter output with adjacent channel signals turned off.
(b) Waveform statistics.

Next the 40-kHz carrier is turned off and the 30- and 50-kHz carriers are turned back on. The total power at the IF filter output is again measured. The power ratio is found to be

$$\frac{\text{Signal Power}}{\text{Interference Power}} = \frac{(.4345)^2}{(.0145)^2} = 897.9, \text{ or } 29.5 \text{ dB} \qquad (1)$$

Further Investigations

The 30-dB selectivity measurement obtained in (1) can be improved upon by changing the IF filter characteristics and/or changing the maximum allowable message bandwidths for the given channel spacing. The sweep rate of the sinusoidal message source also controls the resulting spectral sideband level. As a starting point, consider increasing just the order of the IF filter from 5 to 7 poles.

Chapter 12. FM Quadrature Detector

SystemView File: `com_quad.svu`

Problem Statement

A popular integrated circuit-based FM demodulator is known as a quadrature discriminator or quadrature detector. Before constructing a SystemView simulation, the operation of the quadrature detector will be briefly explained. A simplified block diagram of the quadrature detector is shown below.[2]

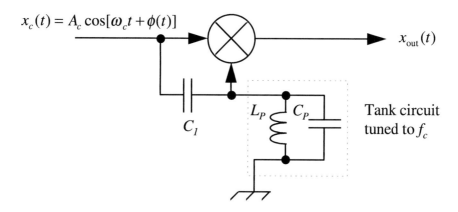

Figure 1: Quadrature detector simplified circuit diagram

The input FM signal connects to one port of a multiplier (product device). The phase deviation function is of the form

$$\phi(t) = K_D \int^t m(\lambda)\, d\lambda \qquad (1)$$

where K_D is the FM modulator deviation constant. A quadrature signal is formed by passing the input to a capacitor series connected to the other multiplier input and a parallel tank circuit resonant at the input carrier frequency.

---

[2] L. W. Couch II, *Modern Communication Systems: Principles and Applications*, Prentice-Hall, Englewood Cliffs, NJ, 1995, p. 271.

- The quadrature circuit receives a phase shift from the capacitor and an additional phase shift from the tank circuit.

- The phase shift produced by the tank circuit is time varying in proportion to the input frequency deviation.

A mathematical model for the circuit begins with the FM input signal

$$x_c(t) = A_c \cos[\omega_c t + \phi(t)] \tag{1}$$

The quadrature signal (second mixer input) is

$$x_{\text{quad}}(t) = K_1 A_c \sin\left[\omega_c t + \phi(t) + K_2 \frac{d\phi(t)}{dt}\right] \tag{2}$$

where the constants K_1 and K_2 are determined by circuit parameters. The multiplier output, assuming a lowpass filter removes the sum terms, is

$$x_{\text{out}}(t) = \frac{1}{2} K_1 A_c^2 \sin\left[K_2 \frac{d\phi}{dt}\right] \tag{3}$$

By proper choice of K_4, the argument of the sin function is small, and a small angle approximation yields

$$x_{\text{out}}(t) \approx \frac{1}{2} K_1 K_2 A_c^2 \frac{d\phi}{dt} = \frac{1}{2} K_1 K_2 A_c^2 K_D m(t) \tag{4}$$

SystemView Simulation

At first glance it might appear that the quadrature detector can only be simulated using a circuit level tool, but as we shall see this is not the case. The series capacitor introduces a phase shift at the carrier frequency.

- One system model for this is a Hilbert transforming filter that introduces the required $90°$ of phase shift across a band of frequencies

- A second model is simply a time delay corresponding to one quarter of a carrier cycle, i.e., $\tau = 1/(4f_c)$

The parallel LC tank circuit can be viewed as a second-order bandpass filter (first-order lowpass prototype) since it has the desired $90°$ of phase shift at the carrier center frequency. The bandwidth of this bandpass filter can be used to set the tank circuit Q and thereby control the parameter K_2. The SystemView simulation block diagram shown below in Figure 2 considers both $90°$ phase-shifter implementations.

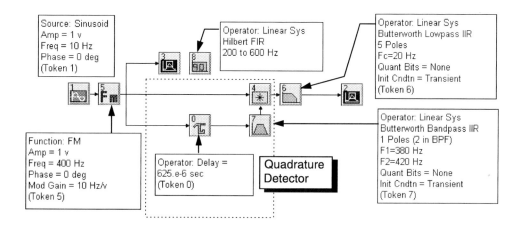

Figure 2: SystemView simulation block diagram contained in file com_quad.svu.

For this experiment, the peak frequency deviation is 10 Hz, the message frequency is 10 Hz, the carrier frequency is 400 Hz, and the sampling frequency is 1600 Hz. Note that a perfect $90°$ phase shift can be obtained by setting the sampling frequency to a multiple of four times the carrier frequency. The maximum message bandwidth is limited to 20 Hz by the lowpass filter used to remove the double-frequency terms from the detector output, but this can easily be changed. The input spectrum is shown below in Figure 3.

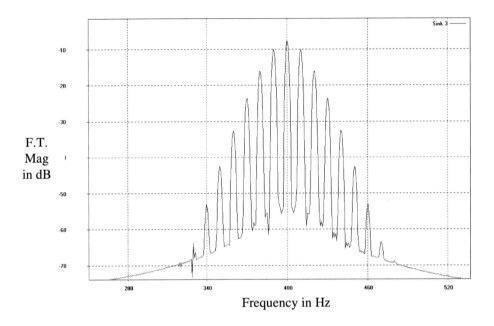

Figure 3: Modulator output spectrum.

The bandpass filter that models the LC tank circuit is designed to have a 20-Hz 3-dB bandwidth. The phase and magnitude response of the LC tank circuit model, as obtained from the SystemView linear system token dialog box, is shown in Figure 4.

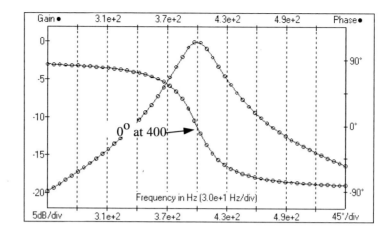

Figure 4: Resonator magnitude and phase response.

By adjusting the bandwidth of the tank circuit (bandpass filter) the distortion level is controllable with respect to the peak deviation level. Ideally, the phase response should be linear in the deviation bandwidth. The detector time-domain output when using the delay line phase shifter is shown in Figure 5.

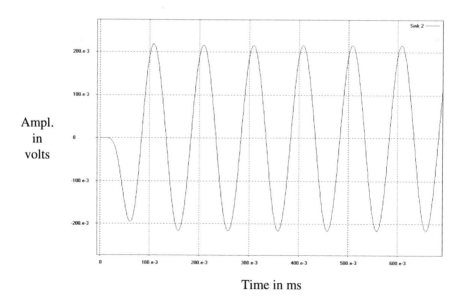

Time in ms

Figure 5: Quadrature detector-time domain output.

The detector frequency-domain output showing harmonic distortion terms is given in Figure 6.

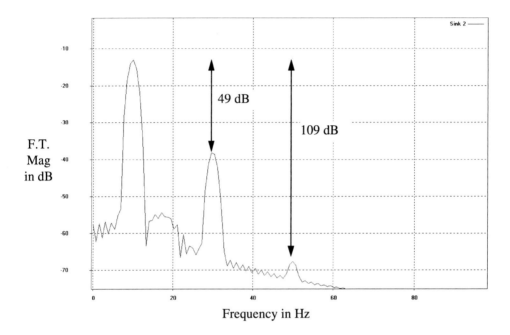

Figure 6: Quadrature detector output spectrum showing harmonic distortion levels.

Similar results are obtained when using the Hilbert phase shifter. The SystemView file contains the Hilbert phase shifter filter as an unconnected token. To try it out, simply disconnect the delay line phase shifter, move the Hilbert phase shifter into position, and make the new connections.

Further Investigations

- Vary the carrier frequency about the nominal design value and observe how rapidly the distortion level increases.
- For a fixed-input frequency deviation, adjust the resonator 3 dB bandwidth to see if lower harmonic distortion can be obtained.
- For the delay line phase shifter vary τ about the nominal value.
- Vary the peak frequency deviation of the modulator.
- Add noise to the input.

Chapter 13. Phase-Locked Loops in Communications

SystemView File: com_pll1.svu, com_pll2.svu, and com_pll3.svu

Problem Statement

In this example, first-order and second-order phase-lock loops are studied in a communication systems context. In particular, the demodulation of frequency modulation (FM) is considered. A basic phase-lock loop (PLL) block diagram is shown in Figure 1.[1]

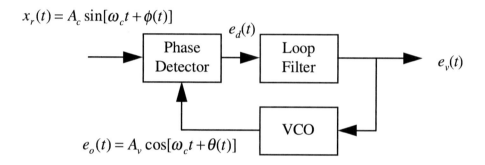

Figure 1: Basic PLL block diagram.

The block labeled VCO in Figure 1 denotes a voltage controlled oscillator, which is a fundamental component in all analog PLLs. In this example, a *sinusoidal phase detector* is assumed. A sinusoidal phase detector is nothing more than an ideal product device. Under the sinusoidal phase detector assumption, an equivalent baseband PLL model, valid for noise-free loop operation, is that shown in Figure 2.

_____

[1] R. Ziemer and W. Tranter, *Principles of Communications*, Fourth Edition, Houghton Mifflin, Boston, MA, 1995, p. 210.

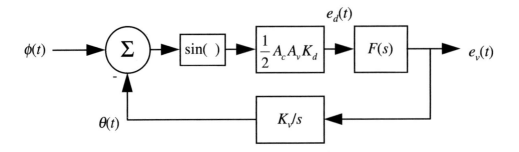

Figure 2: Baseband PLL block diagram.

The describing equation for the baseband PLL model is

$$\frac{d\theta(t)}{dt} = \frac{A_c A_v K_d K_v}{2} \int^t f(t-\alpha)\, \sin[\phi(\alpha) - \theta(\alpha)]d\alpha \quad (1)$$

where K_d is the phase detector gain, K_v is the VCO gain in rad/s/v, and $f(t)$ is the impulse response of the loop filter. The linear PLL model is obtained by removing the sin( ) nonlinearity. In the s-domain, the linear PLL model is as shown in Figure 3.

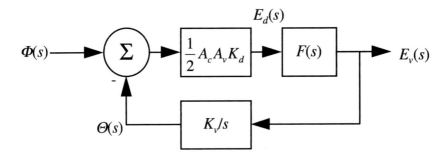

Figure 3: Linear baseband PLL model.

Using the linear model we can now solve for the closed-loop transfer function, $\Theta(s)/\Phi(s)$

$$H(s) = \frac{\Theta(s)}{\Phi(s)} = \frac{KF(s)}{s + KF(s)} \qquad (2)$$

where $K = A_c A_v K_d K_v/2$ and $F(s)$ is the Laplace transform of $f(t)$.

First-Order PLL

Let $F(s) = K_a$, then we have

$$H(s) = \frac{KK_a}{s + KK_a} \qquad (3)$$

An FM input with a message signal of the form $m(t) = A\, u(t)$ (a frequency step) is

$$\phi(t) = Ak_f \int^t u(\alpha)d\alpha \quad \Rightarrow \quad \Phi(s) = \frac{Ak_f}{s^2} \qquad (4)$$

where k_f is the FM modulator deviation constant in Hz/v. The VCO control voltage should be closely related to the applied FM message. To see this, we can write

$$E_v(s) = \frac{s}{K_v}\Theta(s) = \frac{s}{K_v}\Phi(s)H(s) = \frac{Ak_f}{K_v} \cdot \frac{K_t}{s + K_t} \qquad (5)$$

where $K_t = KK_a$. Partial fraction expanding and inverse Laplace transforming yields

$$e_v(t) = \frac{Ak_f}{K_v}(1 - e^{-K_t t})u(t) \qquad (6)$$

Now, in general, if the bandwidth of $m(t)$ is $W \ll K_t$, then

$$E_v(s) \approx \frac{k_f}{K_v} M(s) \implies e_v(t) \approx \frac{k_f}{K_v} m(t) \qquad (7)$$

where $M(s)$ is the Laplace transform of $m(t)$. Note that due to the sin nonlinearity, the first-order PLL has finite *lock range*, K_f, and hence always has a nonzero steadystate phase error when the input frequency is offset from the quiescent VCO frequency. Due to the presence of spurious time constants in the loop, it also very difficult to build a true first-order PLL.

Second-Order Type II PLL

To mitigate some of the problems of the first-order PLL, we can include a second integrator in the open-loop transfer function. A common loop filter for building a second-order PLL consists of an integrator with phase lead compensation, i.e.,

$$F(s) = \frac{s + \tau_2}{s\tau_1} \qquad (8)$$

The resulting PLL is sometimes called a perfect second-order PLL since two integrators are now in the open-loop transfer function. The closed-loop transfer function is of the form

$$H(s) = \frac{2\zeta\omega_n s + \omega_n^2}{s^2 + 2\zeta\omega_n s + \omega_n^2} \qquad (9)$$

where $\omega_n = \sqrt{K/\tau_1}$ and $\zeta = \tau_2\sqrt{K/\tau_1}/2$. For an input frequency step, the steady-state phase error is zero and the loop *hold-in range* is infinite, in theory, since the integrator contained in the loop filter has infinite dc gain.

SystemView Simulation

In SystemView, the simulation can be performed using either with the actual bandpass signals or at baseband using the nonlinear model of Figure 2. The most realistic simulation method is to use the bandpass signals, but since the carrier frequency must be kept low to minimize the simulation time, we have difficulties removing the double frequency term from the phase detector output. By simulating at baseband using the nonlinear loop model, many PLL aspects can be modeled without worrying about how to remove the double frequency term.

First-Order PLL

The SystemView baseband nonlinear PLL simulation block diagram is shown in Figure 4.

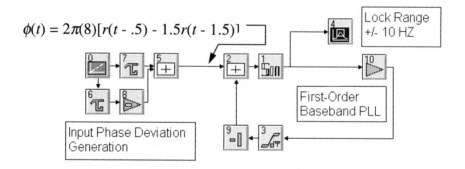

$$\phi(t) = 2\pi(8)[r(t - .5) - 1.5r(t - 1.5)]$$

Lock Range +/- 10 HZ

First-Order Baseband PLL

Input Phase Deviation Generation

Figure 4: SystemView first-order PLL baseband simulation.

To simulate a frequency step input, we input a phase ramp. In this simple example, the input is actually two steps: 8-Hz step turning on at 0.5 seconds and a -12-Hz step turning on at 1.5 seconds. The phase detector output signal (VCO input), $e_d(t)$, is shown in Figure 5.

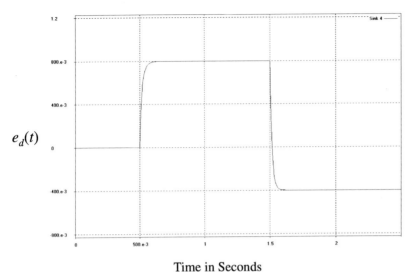

Time in Seconds

Figure 5: Phase detector output $e_d(t)$ in response to a positive and negative step input.

Here we see the finite rise-time due to the loop gain being $K_t = 2\pi(10)$ rad/s. The loop stays in lock since the frequency swing on either side of zero is within the 10-Hz lock range. A detailed plot of the positive step response is shown in Figure 6.

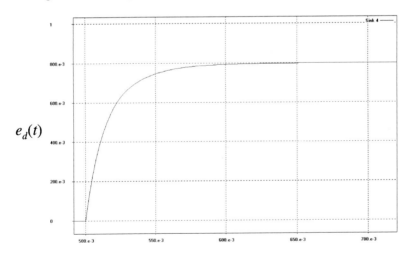

Time in Seconds

Figure 6: A closer look at the phase detector output $e_d(t)$ in response to a positive negative step input.

Suppose now that a single positive frequency step of 12 Hz is applied to the loop. The lock range is exceeded, so the loop unlocks and the cycle slips indefinitely as shown in Figure 7.

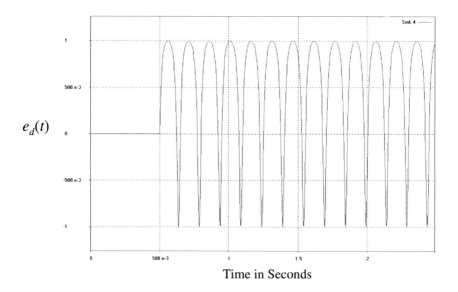

Figure 7: Phase detector output $e_d(t)$ in response to a frequency step of 12 Hz.

Second-Order PLL

As a simulation example, consider a loop designed with $\omega_n/(2\pi) = 10$ Hz and $\zeta = 0.707$. The SystemView baseband simulation block diagram is shown in Figure 8.

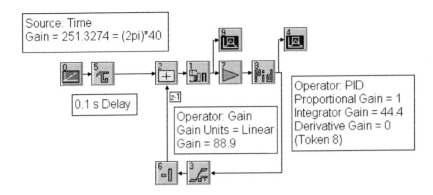

Figure 8: SystemView block diagram for a second-order baseband PLL.

For a 40-Hz input frequency step, the VCO input, which in this case is the VCO frequency offset in rad/s, is shown in Figure 9.

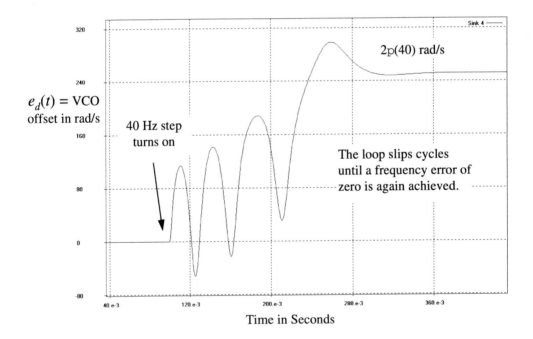

Figure 9: VCO input signal, $e_d(t)$, in response to a 40-Hz frequency step.

Notice that the loop filter block used here is the SystemView PID token, which stands for proportional/integral/derivative feedback control. In SystemView the PID block is of the form

$$F(s) = c_0 + \frac{c_1}{s} + c_2 s \qquad (10)$$

Here we have set $c_2 = 0$ since no derivative control is being used. The PID coefficients are found by equating the two forms for $F(s)$.

A Bandpass Simulation of FM Demodulation

Baseband simulations are very useful and easy to implement, but sometimes a full bandpass-level simulation is required. A simple first-order PLL for FM demodulation is shown in Figure 10.

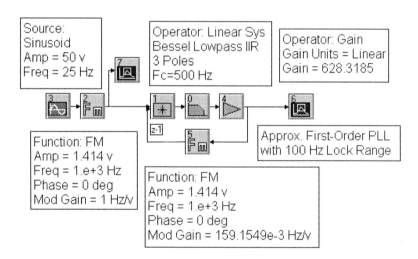

Figure 10: SystemView bandpass simulation of a first-order PLL.

The modulator input is a sinusoid of amplitude 50 at 25 Hz, thus since the FM source has modulation gain of 1 Hz/v, the peak deviation is 50 Hz. The loop lock range is 100 Hz, so the loop should remain in lock; also the closed-loop 3-dB bandwidth is 100 Hz, so the 25-Hz message is within the passband of the loop. The Bessel lowpass filter is used to remove the double frequency term at the output of the multiplier-type phase detector. In reality, the loop is no longer a first-order loop, but the dominant-pole in the closed-loop response is still approximately set by the loop gain. A plot of the VCO control signal is shown in Figure 11.

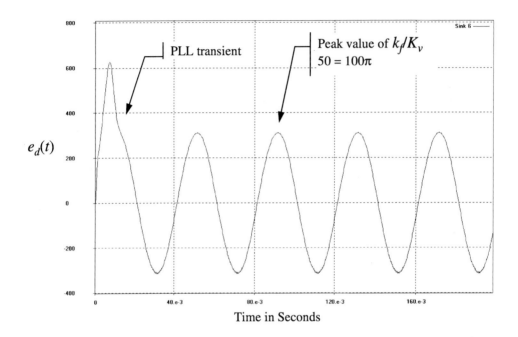

$$e_d(t)$$

Time in Seconds

Figure 11: VCO input signal, $e_d(t)$, in response to 25-Hz FM tone modulation.

After a brief transient, we see that the PLL recovers the FM modulation as expected.

Chapter 14. Baseband Binary Digital Data Transmission

SystemView File: `com_bbb.svu`

Problem Statement

A common starting point in the study of digital communications is a binary signaling scheme that sends constant $\pm A$ amplitude pulses for T_b-second intervals to represent ones and zeros. The mathematical form of the information signal is

$$s(t) = A \sum_{k=-\infty}^{\infty} \alpha_k p(t - kT_b) \qquad (1)$$

where the pulse function $p(t)$ is given by

$$p(t) = \begin{cases} 1, & 0 \le t < T_b \\ 0, & \text{otherwise} \end{cases} \qquad (2)$$

and a_k is a binary symbol modeled as an independent sequence of random variables each equally likely to take on values of ± 1. For testing purposes, the binary data sequence may be generated by a pseudo random-sequence generator (PN sequence)– say, an *m-sequence*. The channel is assumed to corrupt the signal with additive white Gaussian noise (AWGN) with power spectral density of $N_o/2$. The optimum receiver is a synchronous integrate-and-dump detector, as depicted in Figure 1.[2]

_____

[2] R. Ziemer and W. Tranter, *Principles of Communications*, Fourth Edition, Houghton Mifflin, 1995, p. 457.

SystemView Student Edition

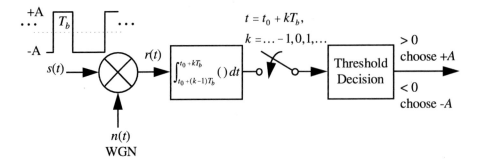

Figure 1: Integrate-and-dump receiver block diagram.

Note t_o in Figure 1 is the synchronization parameter which here is assumed to be known. A useful performance criterion for this receiver is the average probability of bit error. Under the AWGN assumption, this can be shown to be

$$P_E = Q\left(\sqrt{\frac{2A^2 T_b}{N_0}}\right) = Q\left(\sqrt{2E_b/N_0}\right) \qquad (3)$$

where $Q(\ )$ is the Gaussian Q-function defined by

$$Q(x) = \int_x^\infty \frac{1}{\sqrt{2\pi}} e^{-u^2/2} du \qquad (4)$$

and $E_b/N_0 = A^2 T_b/N_0$ is the ratio of energy-per-bit to noise-power density. In binary digital communications, E_b/N_0 is often equated with the signal-to-noise ratio (SNR).

In this example, the binary data transmission scheme described above is simulated so that an experimental verification of the bit-error probability (BER) expression in (3) can be obtained. This simulation will constitute what is known as a *monte-carlo simulation,* in which repeated trials are performed in order to estimate the statistics of one or more system performance measures. Simulation of digital modulation schemes is particularly important when the modulation scheme is complex and/or when the system and channel impairments make analytical calculations very difficult. For the system described here, setting up a simulation is not really required, but it is instructive in learning monte-carlo simulation concepts.

SystemView Simulation

A discrete-time simulation of the integrate-and-dump receiving system of Figure 1 can be implemented very easily with SystemView. Since perfect timing is assumed and the channel bandwidth is assumed infinite, the simulation may be reduced to requiring only one simulation sample per bit (symbol). The obvious advantage of this simplification is reduced simulation time. The SystemView block diagram of a one-sample-per-bit system is shown in Figure 2.

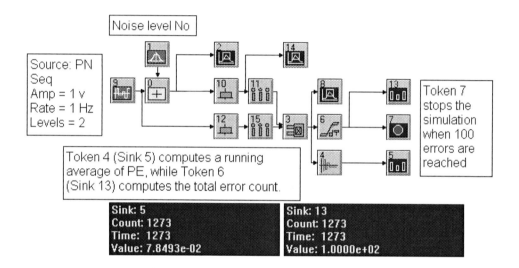

Figure 2: SystemView one-sample-per-bit simulation block diagram.

If increased waveform fidelity is required–if channel bandlimiting is introduced, say, or if a suboptimal detection filter is to be studied–the simulation can be reconfigured by changing the sampling rate. Currently, the bit rate is $R_b = 1$ b/s and the sampling rate is $f_s = 1$ Hz. Increasing the sample rate to 10 Hz provides 10 samples per bit in the simulation. The random data source has amplitude -1, and with only one sample per bit the SNR parameter, E_b/N_0, is set by varying the N_0 value of the Gaussian noise source. In general,

$$\text{SNR}_{dB} = 10\log\left[A^2 T_b/N_0\right]^{\text{here}} = 10\log\left[1/N_0\right] \quad (5)$$

To obtain a particular SNR value in the simulation set,

$$N_0 = 10^{-\text{SNR}_{dB}/10} \qquad\qquad (6)$$

The integrate-and-dump filter is implemented in the discrete-time domain using a one-second-window moving average token followed by a sampler running at 1 Hz. A copy of the transmitter data source is also passed through a similar moving average/sampler token cascade. With both the noisy received and clean transmitter bit streams properly time aligned and sampled, the exclusive-or token then performs hard decisions using a threshold of zero in combination with error detection. The integrator token sums the errors and stops the simulation if 100 error events are reached before the normal stopping time. The running average token computes the probability of error estimate by continuously computing the ratio of error events to total bits processed.

BER Testing

The results of a 0-dB SNR baseline BER test are shown in Figure 3 as a running estimate of P_E.

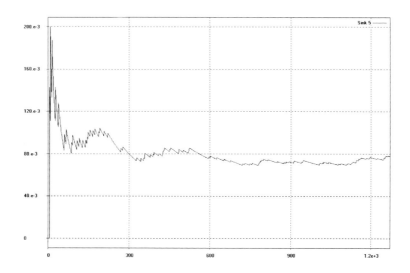

Figure 3: Running average estimate of P_E for 0-dB SNR.

When 100 error events were reached, the simulation stopped, with the final P_E estimate being $\hat{P}_E = 0.07849$ (this result can be seen in the screen capture of Figure 2). The theoretical value is $P_E = 0.07850$. The estimate $\hat{P}_E$ is, of course, a random variable with a mean hopefully the true value of P_E and a variance that decreases as the number of error events increase. Additional simulation results are given in Table 1.

Table 1: BER simulation results for a 100-error threshold

E_b/N_o (dB)	P_E, theory	P_E, experiment
0	7.86E-2	7.72E-2
2	3.75E-2	3.45E-2
4	1.25E-2	1.40E-2
5	5.95E-3	6.77E-3
6	2.39E-3	2.35E-3
7	7.73E-4	9.16E-4 (30 errors in 32766 trials)

The results of Table 1 are plotted in Figure 4 against the known theoretical P_E expression of (3).

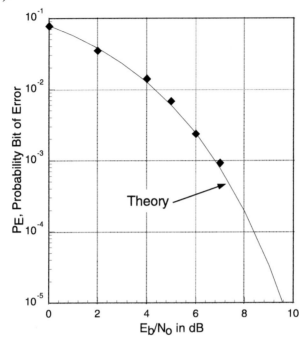

Figure 4: Simulation versus theoretical BER 100-error events.

Eye Patterns

When there is more than one sample per bit we have an actual waveform-level simulation. The receiver matched filter output (moving average filter) can be observed using a SystemView time slice plot, which overlays multiple, fixed time-interval segments of a sink token data record. For the matched filter output, the time interval should be an integer multiple of the bit period; here, that is one second. Digital communication engineers refer to these plots as *eye patterns* since at the sampling instant the overlay of all the plots has an opening, and half a bit period either side of this opening is where the bit transitions occur. The shape of the opening thus looks similar to an eye. When noise is present, the opening closes somewhat, depending upon the SNR level.

In Figures 5 and 6 eye patterns are shown for 10 samples per symbol with no noise and a 10-dB SNR, respectively.

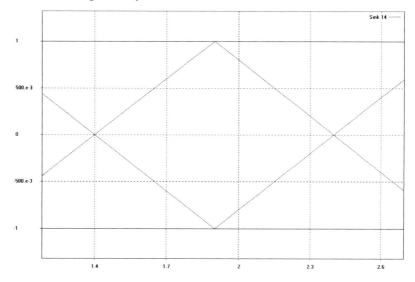

Figure 5: Eye pattern plot at matched filter output with no noise and 10 samples per bit.

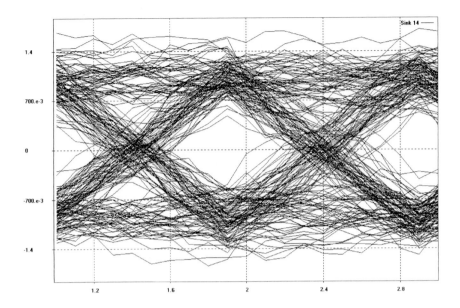

Figure 6: Eye pattern plot at matched filter output with a 10-dB SNR and 10 samples per bit.

At a 10-dB SNR the eye shown in Figure 6 is still open and this is expected since $P_E = 3.87\text{E-}6$.

Further Investigations

At present, the channel is distortionless. With a the sampling rate at 10 samples per bit or so, a lowpass filter may be placed between the transmitter and receiver to model bandwidth limitations imposed by a physical channel. The sample time may also be skewed from the point of maximum eye opening to represent a static timing error degradation. If a lowpass filter is inserted in the channel, a delay token will be needed ahead of the sampler to compensate for delay introduced by the filter. To ensure the reference bit stream from the transmitter is aligned properly with the receiver data, a similar delay must also be placed in front of token 15. For more bandwidth-efficient baseband communication, square-root-raised cosine filtering at the transmitter and the receiver may be implemented. In this case, the receiver moving average filter would be replaced by one of the square-root-raised cosine filters.

Chapter 15. A BPSK Modem with Adjacent Channel Interference

SystemView File: `com_psk.svu`

Problem Statement

In this example, a binary phase-shift keyed (BPSK) modem is simulated in a practical multiuser environment. The modem block diagram is shown in Figure 1.

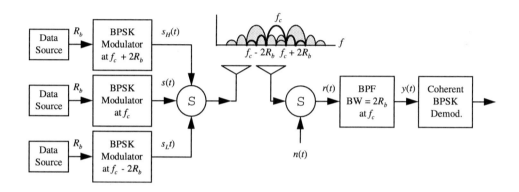

Figure 1: Modem block diagram including adjacent channel signals.

The receiver is subject to adjacent channel interference (ACI) caused by like-data-rate BPSK carriers spaced in frequency at either side of the signal of interest. The carrier spacing is nominally twice the bit rate. To minimize the ACI, a 5-pole Chebyshev bandpass filter with a 3-dB bandwidth equal to the BPSK mainlobe bandwidth is placed at the receiver input. The bandpass filter must remove as much ACI as possible, yet cause minimal intersymbol interference (ISI). Additive white Gaussian noise (AWGN) is also present. It is assumed that all three signals have the same carrier power.

A detailed analysis of this system to determine the average probability of bit error is far from trivial. Simulation, on the other hand, is rather simple, particularly with a tool such as SystemView. As a very simple analysis approach, start with the ideal AWGN BPSK probability of bit error (BER) expression and modify the expression for energy per bit, E_b, to include energy loss due to bandlimiting and noise power spectral density, N_o to include increased noise due to ACI. For ideal BPSK, the BER expression is

$$P_E = Q\left(\sqrt{2\,E_b/N_o}\right) = Q\left(\sqrt{2 \cdot \text{SNR}}\right) \tag{1}$$

where Q is the Gaussian Q-function and SNR = E_b/N_o.[3] To account for signal energy loss due to bandlimiting, we let

$$E_b' = E_b \frac{\int_{-1}^{1} \text{sinc}^2(f)df}{\int_{-\infty}^{\infty} \text{sinc}^2(f)df} = K_1 E_b \tag{2}$$

To account for increased noise power due to ACI, we let

$$N_o' = N_o + 2E_b \tfrac{1}{2}\int_{1}^{3} \text{sinc}^2(f)df = N_o + 2K_2 E_b \tag{3}$$

Numerically evaluating the integrals in (2) and (3) results in $K_1 = 0.903$ and $K_2 = 0.0159$. Finally, the BER expression becomes

$$P_E' = Q\left(\sqrt{\frac{2K_1 E_b}{N_o + 2K_2 E_b}}\right) = Q\left(\sqrt{\frac{2K_1 \cdot \text{SNR}}{1 + 2K_2 \cdot \text{SNR}}}\right) \tag{4}$$

Note that by letting $K_2 = 0$ in (4), the model considers just degradation due to ISI.

---

[3] R. Ziemer and W. Tranter, *Principles of Communications*, Fourth Edition, Houghton Mifflin, Boston, MA, 1995, p. 477.

SystemView Student Edition

SystemView Simulation

The SystemView simulation block diagram is shown in Figure 3.

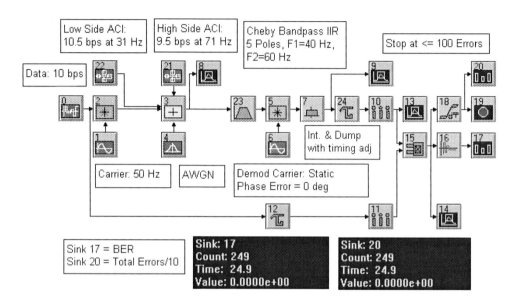

Figure 3: SystemView simulation block diagram of a
BPSK modem that includes ACI, ISI, and AWGN.

The communication parameters in the simulation have values that are in proportion to the block diagram of Figure 1. In particular, R_b = 10 b/s, f_c = 50 Hz, and the simulation sampling rate is 200 Hz. To give a more asynchronous like characteristic to the ACI, the low-side BPSK signal is centered at 31 Hz with a data rate of 10.5 b/s, and the high-side BPSK signal is centered at 71 Hz with a data rate of 9.5 b/s. The bandpass filter has 3-dB frequencies at 40 and 60 Hz. The sampling rate of 200 Hz was chosen as a compromise between minimizing aliasing effect yet trying to keep the simulation efficient in terms of CPU time. As it stands, each bit is represented with 200/10 = 20 samples. The details of the error detection scheme and P_E calculation are explained more fully in the SystemView example *Baseband Binary Digital Data Transmission*. The received noise-free signal spectrum is shown in Figure 3 using a single periodogram to estimate the spectrum.

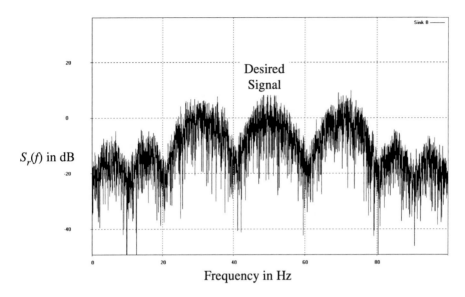

Figure 3: Received spectrum with the noise turned off.

Eye Pattern Results

To observe the influence of the bandpass filter on just the signal component, eye pattern plots with and without the ACI turned are shown in Figures 4 and 5, respectively.

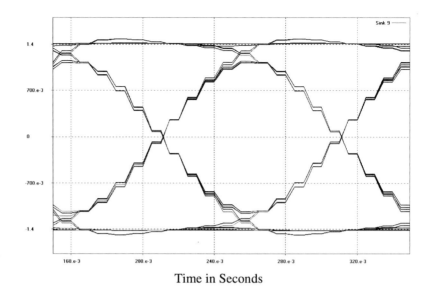

Time in Seconds

Figure 4: Matched filter output eye pattern with no ACI present.

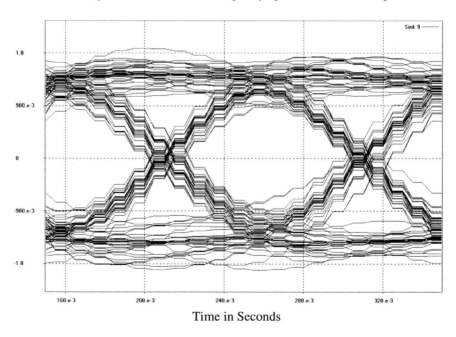

Time in Seconds

Figure 5: Matched filter output eye pattern with ACI present.

The *stair-steps* in both Figures 4 and 5 are due to the fact that the double frequency term present in the demodulation multiplier output is not fully suppressed by the moving average filter. With no ACI present, we see that the eye is about 80% open, while with ACI also present the eye is only 60% open.

BER Results

BER results were obtained by running the simulation until 100 error events occurred. The noise level, N_o, is set using the formula

$$N_o = 10^{-\left(\frac{SNR_{dB}}{10}+1\right)} \tag{5}$$

The +1 term in the exponent accounts for the fact that $R_b = 10$ b/s. Simulation results were first obtained with the ACI turned off then with the ACI turned on. In both cases, ISI is present.

Table 1: BER simulation results for a 100-error threshold.

E_b/N_o (dB)	P_E, ideal theory	P_E, ISI only model	P_E, ACI/ISI model	P_E, ISI experiment	P_E, ACI/ISI experiment
0	7.86E-2	8.95E-2	9.29E-2	9.37E-2	9.60E-2
2	3.75E-2	4.53E-2	4.94E-2	4.08E-2	5.78E-2
4	1.25E-2	1.66E-2	2.02E-2	2.06E-2	2.18E-2
5	5.95E-3	8.43E-3	1.14E-2	9.53E-3	1.04E-2
6	2.39E-3	3.67E-3	5.76E-3	3.88E-3	6.47E-3

The results of Table 1 are plotted below in Figure 6.

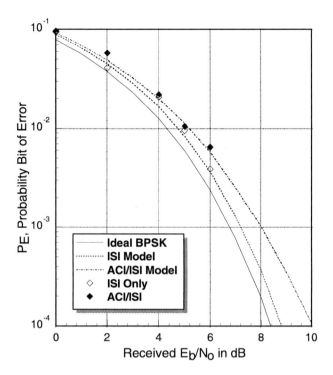

Figure 6: BER simulation results versus ideal theory
and the simple ACI/ISI model using a 100-error events threshold.

The simulation results and the simple ACI/ISI model agree fairly well at the E_b/N_o tested. To get lower BER values long simulation times can be expected.

Further Investigations

Keeping the same theme of degradation due to ACI and ISI, a different receiver bandpass filter may be used, the channel spacing may increased, and shaping of the transmitted BPSK spectra may be considered. Additional degradation results to consider are static-timing error in at the matched filter sampler and static-carrier-phase error in at the demodulation multiplier.

Chapter 16. Direct Sequence Spread-Spectrum with Noise and Tone Jamming

SystemView File: `com_dsss.svu`

Problem Statement

Direct-sequence spread-spectrum (DSSS) is a modulation technique that spreads the transmitted signal bandwidth so that it is much greater than the inherent bandwidth of the modulating signal. DSSS finds application in military communication systems as well as commercial multiple access communication systems. A basic DSSS transceiver system is shown in Figure 1.

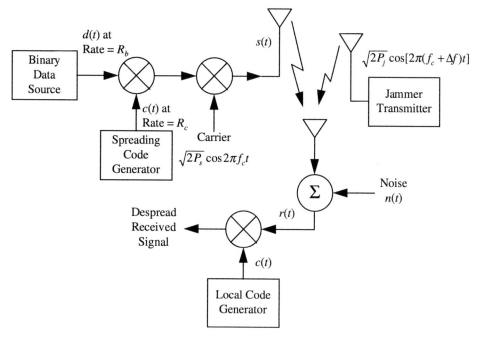

Figure 1: DSSS transceiver block diagram.

The channel model indicated in Figure 1 consists of additive white Gaussian noise (AWGN) and a single frequency tone jammer. Many other channel scenarios are appropriate for DSSS system analysis as well. The primary intent of this SystemView

SystemView Student Version

example is to show via spectral plots the processing gain DSSS offers over unspread modulation, when a narrowband jammer is present. Secondly, the ability to hide a spread signal in noise will also be observed. The signal time- and frequency-domain models for Figure 1 are know briefly investigated.[1]

For binary phase-shift-keyed (BPSK) DSSS, the transmitted signal is of the form

$$s(t) = \sqrt{2P_s} d(t) c(t) \cos(2\pi f_c t + \theta) \tag{1}$$

where P_s is the signal power, $d(t)$ is a ± 1 bit sequence with bit duration T_c, $c(t)$ is the spreading code sequence which is a ± 1 binary sequence with *chip* duration T_c, and f_c is the carrier frequency. The carrier phase is assumed to be uniformly distributed in $[0, 2\pi)$. Assuming independence between $d(t)$ and $c(t)$ and each being random binary sequences, we can write that the power spectrum of $s(t)$ is

$$S_s(f) = \frac{P_s T_c}{2} \left\{ \text{sinc}^2 [T_c(f - f_c)] + \text{sinc}^2 [T_c(f + f_c)] \right\} \tag{2}$$

where $\text{sinc}(v) = \sin(\pi v)/(\pi v)$. The received signal (neglecting propagation delays) is of the form

$$r(t) = s(t) + n(t) + \sqrt{2P_j} \cos[2\pi(f_c + \Delta f) + \phi] \tag{3}$$

where $n(t)$ is AWGN of spectral density $N_o/2$, P_j is the jammer power, Δf is the jammer frequency offset from the carrier, and ϕ is independent and uniformly distributed in $[0, 2\pi)$. The received signal power spectrum is of the form

---

[1] R. Ziemer and W. Tranter, *Principles of Communications*, Fourth Edition, Houghton Mifflin, Boston, MA, 1995, p. 573.

$$S_r(f) = \frac{P_s T_c}{2} \left\{ \text{sinc}^2[T_c(f - f_c)] + \text{sinc}^2[T_c(f + f_c)] \right\} + \frac{N_o}{2} \tag{4}$$
$$+ \frac{P_j}{2} \left\{ \delta(f - f_c - \Delta f) + \delta(f + f_c + \Delta f) \right\}$$

Note that the mainlobe bandwidth of the received signal component is $2/T_c$ embedded in a flat noise spectrum along with a single spectral line component from the jammer.

Following despreading with an identical spreading code sequence, we have

$$z(t) = \sqrt{2P_s} d(t) \cos(2\pi f_c t + \theta) + n(t)c(t)$$
$$+ \sqrt{2P_j} c(t) \cos[2\pi(f_c + \Delta f) + \phi] \tag{5}$$

The corresponding despread signal power spectrum is

$$S_z(f) = \frac{P_s T_b}{2} \left\{ \text{sinc}^2[T_b(f - f_c)] + \text{sinc}^2[T_b(f + f_c)] \right\} + \frac{N_o}{2} \tag{6}$$
$$+ \frac{P_j}{2} \left\{ \text{sinc}^2[T_c(f - f_c - \Delta f)] + \text{sinc}^2[T_c(f + f_c + \Delta f)] \right\}$$

The despread received signal consists of a data spectrum with a mainlobe bandwidth of $2/T_b$, a flat noise spectrum of height $N_o/2$, along with a spread jammer component with a mainlobe bandwidth of $2/T_c$.

SystemView Simulation

The focus of this simulation is on the spectral analysis of the spread and despread received DSSS signal corrupted by AWGN and a single-tone jammer. The simulation block diagram is shown in Figure 2.

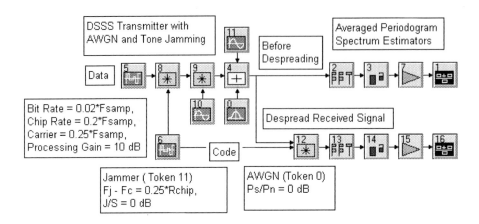

Figure 2: SystemView simulation block diagram.

The spectral estimation capabilities of the SystemView analysis window is limited to FFT analysis of a single-windowed data record. To reduce the variance in the power spectrum estimation of random signals, the *averaged periodogram* is useful. Additional details on this spectral estimation technique can be found in the SystemView example entitled *Averaged Periodogram Spectral Estimation*. Note that a dedicated periodogram algorithm is used to obtain the power spectrum of both the spread and despread signals. The number of periodograms averaged is 25, with the length of each subrecord 512 samples. When the contents of sink tokens 1 and 16 are viewed in the analysis window, the last 257 points, indexed from 0 to 256, contain the desired frequency-domain data. The bin at zero corresponds to dc and the bin at 256 corresponds to $f_s/2$, which here is 1/2 Hz.

Given the 1-Hz sampling rate, the DSSS system parameters are a bit rate of $R_b = 0.02$, chip rate $R_c = 0.2$, $f_c = 0.25$, $\Delta f = 0.2$, $P_J/P_s = 0$ dB, and $P_s/\sigma_n^2 = 0$ dB. The *processing gain* for this system, computed as the ratio of chip rate to bit rate, is

$$G_p = \frac{R_c}{R_b} = \frac{0.2}{0.02} = 10 \text{ or } 10 \text{ dB} \qquad (7)$$

In a single-tone jamming environment, the processing gain is a measure of the jammer suppression resulting from the signal spreading. From a practical standpoint, a processing gain of 10 dB may not be worth the effort. For the purposes of this example problem the desired results are still obtained and without excessive simulation time. Verification of the 10-dB processing gain, in an approximate way, can be seen in the pre- and post-despread power spectra shown in Figures 3 and 4, respectively.

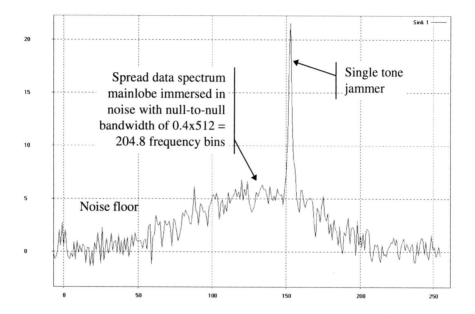

Figure 3: Received DSSS signal prior to despreading with noise and a single-tone jammer present.

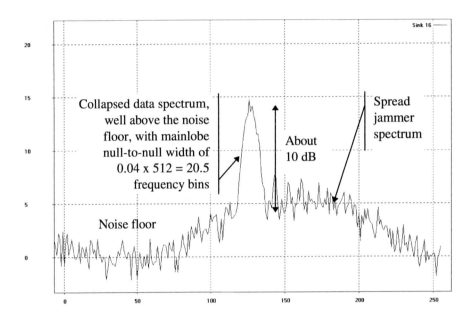

Figure 4: Received DSSS signal following despreading
with noise and a single-tone jammer present.

In comparing Figure 3 and Figure 4 we also see that prior to despreading the signal is almost totally immersed in the additive noise. Following the despreader, the signal spectrum is compressed by a factor of 10, and hence the spectral density increases allowing the mainlobe to push up out of the noise and the now-spread jammer spectrum.

Further Investigations

The DSSS simulation example as presented here serves as a starting point for additional DSSS communication system investigations. To begin with, create a copy of the file com_dsss.svu. Remove the spectral estimation tokens and replace them with BPSK data demodulation functions. A complete BPSK data modem is contained in the example file com_psk.svu. Investigate the *eye diagram* produced at the matched filter output for different values of P_j/P_s and Δf. In SystemView, eye diagrams are produced using the slicing option under the Preferences menu. Examples of eye diagrams are contained in the BPSK modem example. With Δf within the bit-rate bandwidth, investigate the eye pattern with and without spreading/despreading. As a follow-up to the eye diagram study, bit-error detection can then be added so that the bit-error rate can be estimated for a least high P_j/P_s ratios.

The simulation fidelity could also be improved by increasing the ratio of the sampling rate to chip rate. This, of course, will increase the time required to simulate each data bit. The number of samples required to simulate one data bit becomes very costly when the processing gain is increased to more realistic values.

Chapter 17. Lowpass and Bandpass Sampling Theory Application

SystemView File: `dsp_samp.svu`

Problem Statement

In this example we use SystemView to examine both lowpass sampling and bandpass sampling. The system block diagram is given in Figure 1 below.

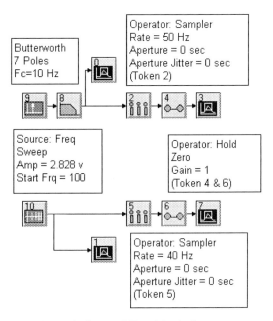

Figure 1: SystemView block diagram.

The input lowpass spectrum is created from the impulse response of a seventh-order lowpass filter with $f_c = 10$ Hz. From the lowpass sampling theorem,

$$f_s > 2 \times 10 = 20 \text{ Hz} \quad (1)$$

The bandpass signal is created by sweeping a sinusoid from 100 Hz to 120 Hz in 2 s. The bandpass sampling theorem allows[2]

$$f_s = \frac{2 \times 120}{[120/20]} = \frac{240}{6} = 40 \text{ Hz} \tag{2}$$

The lowpass and bandpass spectra are shown below in Figures 2 and 3, respectively.

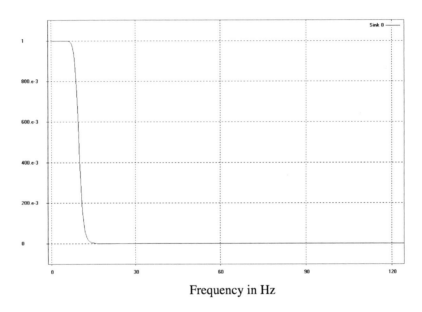

Frequency in Hz

Figure 2: Lowpass spectra prior to sampling.

[2] R. Ziemer and W. Tranter, *Principles of Communications*, Fourth Edition, Houghton Mifflin, Boston, MA, 1995, p. 93.

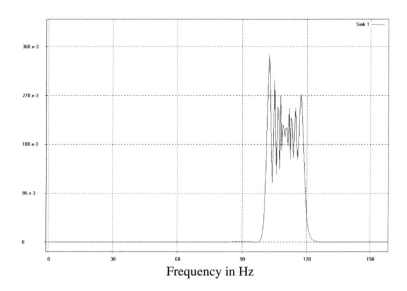

Figure 3: Bandpass spectra prior to sampling.

The lowpass signal is now sampled with $f_s = 50$ Hz; the bandpass signal is undersampled with $f_s = 40$ Hz. Figure 4 shows the sampled lowpass signal spectra.

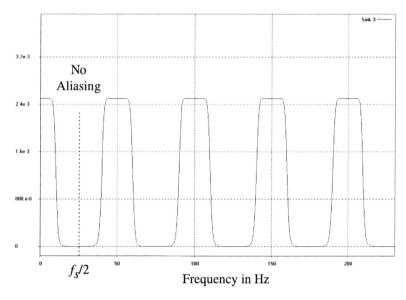

Figure 4: 50-Hz sampled lowpass signal spectrum indicating little or no aliasing.

Note that spectral translates are now located every 50 Hz, but aliasing is not a problem since the lowpass sampling theorem is satisfied. Finally shown in Figure 5 is the spectrum of the undersampled bandpass signal.

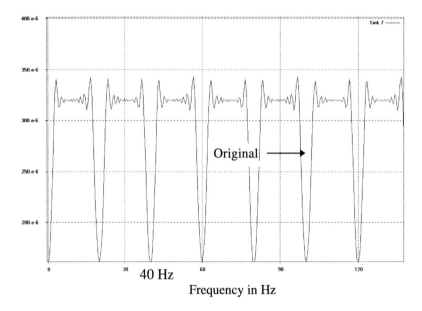

Figure 5: 40-Hz undersampled bandpass signal spectrum showing spectral translates filling in the interval below 100 Hz.

Note that spectral translates are now located every 40 Hz, but aliasing is not a problem since the bandpass sampling theorem is satisfied.

Chapter 18. A Digital Filtering Application

SystemView File: `dsp_fil1.svu`

Problem Statement

The application of interest here is the discrete-time processing of continuous-time signals, i.e., the C/D–$H(e^{j\omega})$–D/C system, where C/D denotes a continuous- to discrete-time conversion operation and D/C denotes a discrete- to continuous-time conversion. SystemView is used to implement a digital filtering operation with a 1000 Hz sampling rate.

Filter Design and Analysis

The heart of this system is a fifth-order Chebyshev lowpass filter that is designed to have a cutoff frequency of 100 Hz, a unity passband gain, and a passband ripple of 0.5 dB. The discrete-time equivalent filter is designed using SystemView's infinite impulse response (IIR) filter design library, which is contained within the linear system operator token (green tokens). The IIR filter design is obtained from an analog prototype using the bilinear transformation. The filter is realized in a direct form structure with a system function of the form

$$H(z) = \frac{\displaystyle\sum_{k=0}^{M} b_k z^{-k}}{\displaystyle\sum_{k=0}^{N} a_k z^{-k}} \tag{1}$$

and a difference equation of the form

$$y[n] = \sum_{k=0}^{M} b_k x[n-k] - \sum_{k=1}^{N} a_k y[n-k] \tag{2}$$

SystemView Student Version

- To start the filter design process, we place a linear system token in the workspace. The entry point into the filter design utility is the first-level dialog box under the linear system token, as shown below in Figure 1.

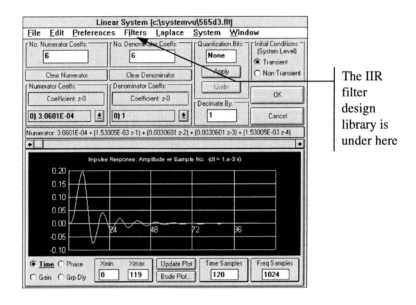

Figure 1: Linear system token main dialog box.

To design the required Chebyshev lowpass filter, bring up the IIR filter library dialog box of Figure 2 by clicking the IIR menu item.

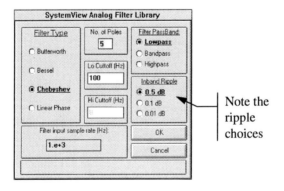

Figure 2: IIR filter design library dialog box.

Once the filter design is chosen and we return to the Figure 1 dialog box, the filter impulse response and frequency response characteristics are immediately available. The direct-form filter coefficients can be read directly from the numerator (b_k's) and denominator (a_k's) drop-down list boxes. The resulting filter coefficients are given in Table 1.

Table 1: Filter coefficients.

Numerator Coefficients =6	Pole Coefficients =6
b0 = 3.0601E-04	a0 = 1
b1 = 1.53005E-03	a1 = -3.90739738361339
b2 = 0.0030601	a2 = 6.48842446406175
b3 = 0.0030601	a3 = -5.67117041404226
b4 = 1.53005E-03	a4 = 2.59848269498342
b5 = 3.0601E-04	a5 = -0.498519929200513

A pole-zero map of the IIR filter can be obtained by choosing the root locus option under the System menu of the linear system dialog box. A root-locus plot is inherently more than simply plotting the poles and zeros of a system function, but here this function is used to plot open-loop poles and zeros a of single-system function. Since the system is discrete, the z-domain option is chosen, and since closed-loop feedback analysis of the filter is not desired, we set the start gain to zero (no feedback) and the stop gain to some small number (0.001) so that just the open-loop poles and zeros of the filter will be displayed, as shown below in Figure 3. Note SystemView requires that at least two gain values be used in the root-locus evaluation.

SystemView Student Version

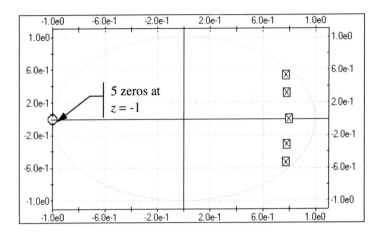

Figure 3: Lowpass filter pole-zero map obtained using root locus with no feedback.

- Note: The five zeros at infinity in the *s*-domain Chebyshev prototype are located at $z = -1$ as expected for a bilinear transformation-based design.

The impulse response of the filter as obtained from the linear system dialog box is shown in Figure 4.

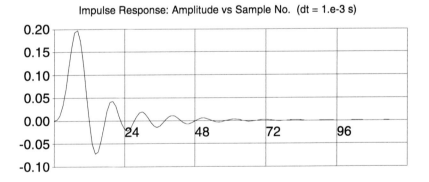

Figure 4: Filter impulse response
(actually, a discrete-time plot, but here the "dots" are connected).

By Fourier-transforming the impulse response, the frequency response can be obtained (in SystemView a fast Fourier transform is used with a user-defined number of points

2^{nu}). Since the frequency response is a complex quantity, we typically view it in terms of magnitude and phase, but in SystemView the *group delay* is also available. Figures 5 - 7 show the filter magnitude response in dB, the unwrapped phase response, and the group delay in samples, respectively. In all plots the frequency axis is a normalized frequency, i.e.,

$$f = \frac{\omega}{2\pi} \text{ or } f = \frac{F_{analog}}{F_{sampling}} \qquad (3)$$

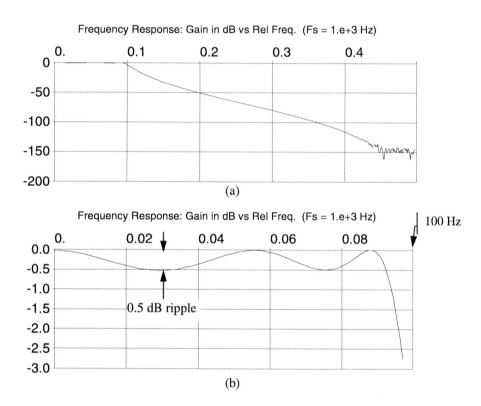

Figure 5: Filter magnitude frequency response versus normalized frequency showing (a) the stopband and (b) a zoomed view of the passband.

SystemView Student Version

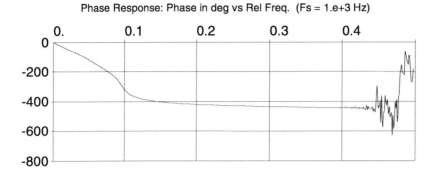

Figure 6: Unwrapped phase versus normalized frequency (the noise-like response is due to numerical problems with angle computation at high attenuation levels).

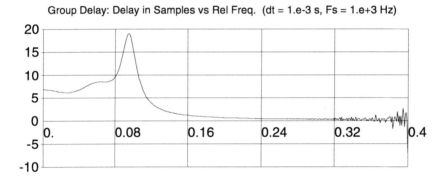

Figure 7: Group delay in samples versus the normalized frequency.

Filtering Signals

Returning now to the C/D–$H(e^{j\omega})$–D/C filter system, we will construct a simple SystemView simulation using the Chebyshev filter designed above. Consider passing several signal types through the filter and observe the outputs. The SystemView simulation block diagram for this experiment is shown in Figure 8.

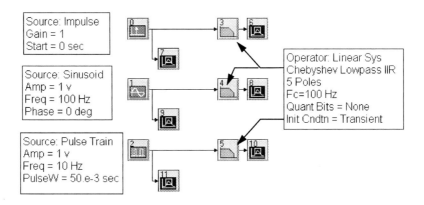

Figure 8: SystemView simulation block diagram
showing the filter in three source/sink networks.

The first system, denoted (a), simply verifies the impulse response as obtained within the token's built-in analysis features. By processing the output with the full output capability of SystemView, a more detailed time- and frequency-domain analysis can be performed. In Figure 9, detailed magnitude and phase plots are given with the graphics cursors used to find the filter gain and phase at the design point of $F_c = 100$ Hz ($f_c = 100/1000 = 0.1$).

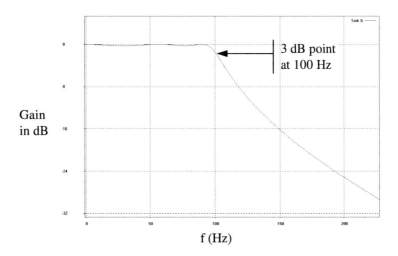

f (Hz)

Figure 9: Filter magnitude response in dB
versus analog frequency with a marker at the 100-Hz point.

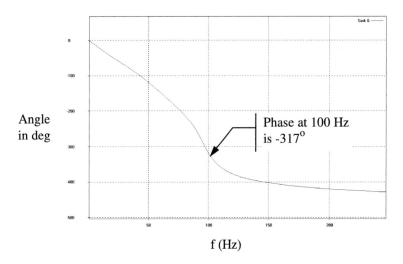

Angle
in deg

Phase at 100 Hz
is -317°

f (Hz)

Figure 10: Filter phase response in degrees
versus analog frequency with a marker at the 100-Hz point.

We will now verify the above magnitude and phase response in the time domain using
system (b), which has as input the sequence

$$x[n] = \cos(2\pi \cdot 100t)\Big|_{t \to \frac{n}{1000}} = \cos\left[2\pi \cdot \frac{100}{1000} \cdot n\right] \qquad (4)$$

From linear system theory we know that the steady-state output, $y[n]$, must be of the
form

$$y[n] = \left|H(e^{j2\pi/10})\right| \cos[2\pi n/10 + \angle H(e^{j2\pi/10})] \qquad (5)$$

Using the SystemView graphics display, as shown in Figure 11, we measure the
magnitude and phase at 100 Hz.

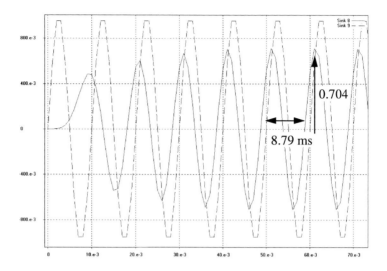

Figure 11: Filter time-domain response to a 100-Hz sinusoid.

The time-domain results obtained in Figure 11 compare favorably with the direct frequency-domain results by noting that

- The steady-state output amplitude is 0.704, thus the filter attenuation at 100 Hz is 1.42 or 3.05 dB.

- The steady-state zero crossing differential is 8.79 ms, which at 100 Hz corresponds to a phase shift of -316.4°.

.

It is also worth noting that in SystemView the cutoff frequency in Chebyshev designs corresponds to the 3-dB bandwidth as opposed to the ripple bandwidth (here 0.5 dB) encountered in other filter design packages. The conversion from ripple bandwidth to a 3-dB bandwidth is, however, trivial in this case. Next, using system (c) the system response to a 10-Hz squarewave input is observed, as given in Figure 12.

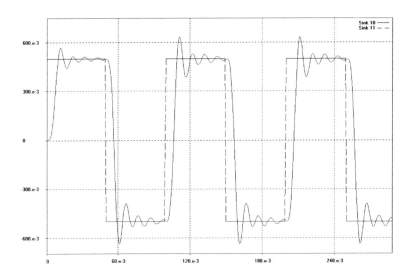

Figure 12:10-Hz squarewave time-domain response.

The filter allows the squarewave harmonics at 10, 30, 50, 70, and 90 Hz to pass through, while the others are greatly attenuated. The fact that the ears on the output are asymmetrical is due to the nonconstant group delay of the filter. From Figure 12 we also observe that steady-state is reached in about one cycle.

Chapter 19. A Multirate Sampling Application

SystemView File: dsp_mult.svu

Problem Statement

In this example we examine a digital audio system that uses both downsampling (decimation) and upsampling (interpolation). The impact of decimation and interpolation is studied in the frequency domain. A secondary issue is the impact of the zero-order hold (ZOH) operation typically found in digital-to-analog (D/A) converters. The basic system model is shown in Figure 1.

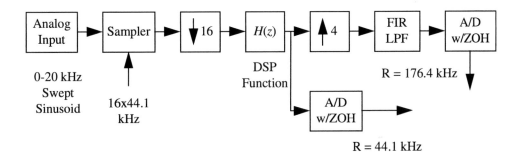

Figure 1: Multirate system block diagram.

The input signal is a swept sinusoid covering 0 - 20 kHz to represent a typical hi-fi audio-type signal. Note that the signal is not strictly bandlimited since the sweep time is finite. The signal is oversampled at a rate approximately 16 times the Nyquist rate so that

- Decimation can be studied without introducing excessive aliasing.

- The spectra of the oversampled signal (actually the base rate for the simulation) can be viewed as an approximation to a continuous-time signal in the simulation.

The SystemView system block diagram is shown in Figure 2.

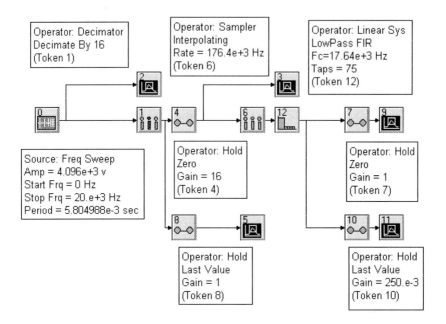

Figure 2: SystemView block diagram.

For this example the discrete-time system, denoted $H(z)$ in Figure 1, is assumed to be a unity gain buffer to more clearly study the influences of the upsampling, down sampling, and the ZOH filter associated with D/A conversion.

System Waveforms and Spectra

Waveforms and signal spectra taken at various points in the simulation block diagram are shown in the following figures. To begin with, Figure 3 is a time-domain plot of the swept "analog" sinusoid.

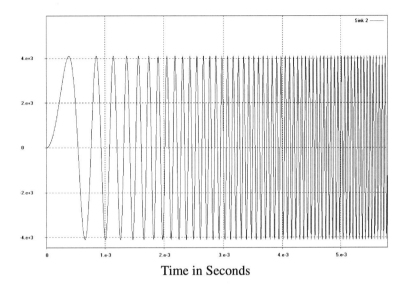

Time in Seconds

Figure 3: Input sinusoid swept from 0 to 20 kHz in 4096 samples
when sampled at rate 16*44.1 kHz or in a time interval of 5.80 ms.

In reality, the signal plotted in Figure 3 is a discrete-time signal, but the sampling
rate is 16 times the Nyquist rate. The spectrum of this signal is shown in Figure 4.

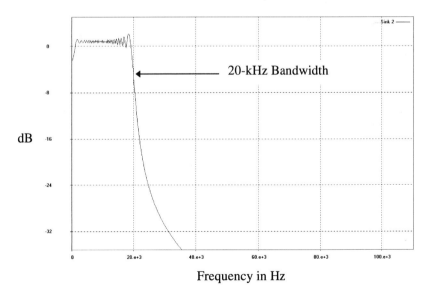

Frequency in Hz

Figure 4: Input signal spectrum in dB versus analog frequency.

Since SystemView only plots the spectrum from zero to $f_s/2$, the first spectral translate of the 16-times oversampled signal centered at 705.6 kHz is not visible. Following the decimation-by-16 operation of token 1 and then the hold zero between samples of token 4, we obtain a signal that resembles ideal impulse train sampling. The spectrum of this signal is shown in Figure 5.

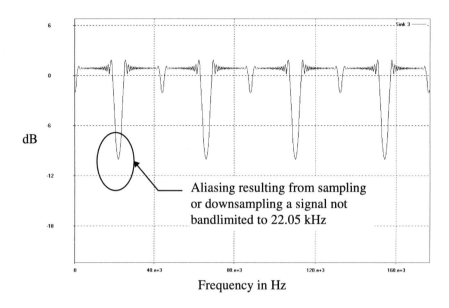

Figure 5: Spectrum in dB following decimation-by-16 or, equivalently, the spectrum of the analog signal sampled at a rate of 44.1 kHz.

This spectrum appears as if the underlying signal is a 44.1-kHz ideal sampled version of the analog signal depicted in Figure 3. If we hold the last sample value as opposed to holding zero between samples, as in token 8, the signal now approximates the output of a D/A converter clocked at 44.1 kHz. The inherent ZOH filter results in the spectral droop seen in Figure 6 below.

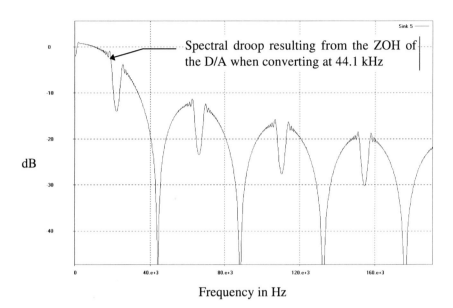

Figure 6: 44.1 kHz D/A output spectrum showing sin(x)/x weighting .

In a compact disk (CD) digital audio playback system with D/A running at 44.1 kHz, the analog reconstruction filter must remove the images above 20 kHz and compensate for the droop caused by the D/A zero-order hold. By upsampling prior to D/A conversion, the analog reconstruction filter requirements can be relaxed. The major portion of the difficult reconstruction filtering task is now performed by the upsampling interpolation filter (token 12). Here this filter is an equal-ripple FIR of length 75 designed to have a 0.1-dB passband ripple and a stopband gain of -35 dB, as shown in Figure 7.

142

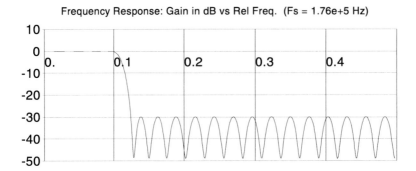

Figure 7: Frequency response of the 75-tap equiripple FIR lowpass filter used to remove spectral images resulting from rate changing from 44.1 kHz to 176.4 kHz (upsampling by 4).

Following the interpolation filter, the spectral images are now located at multiples of 176.4 kHz, and the D/A zero-order hold has its first null at 176.4 kHz, as well. The analog reconstruction filtering is now less critical. The spectrum prior to the D/A ZOH filter is shown in Figure 8.

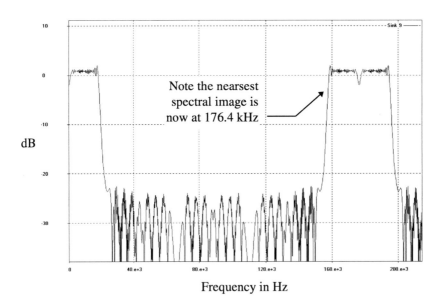

Figure 8: Spectrum in dB seen at the output of the lowpass interpolation filter.

The final output signal spectrum, which includes the ZOH filter, is shown in Figure 9.

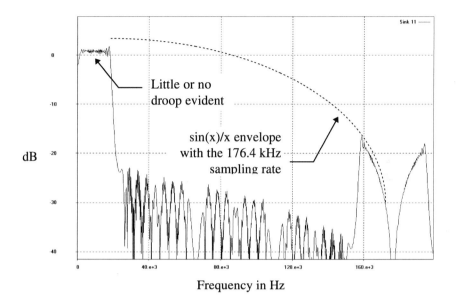

Figure 9: 176.4 kHz D/A output spectrum with reduced droop.

In consumer CD players, upsampling (oversampling in the consumer literature) by four or eight is commonplace.

Chapter 20. FFT Spectral Estimation of Deterministic Signals

SystemView File: `dsp_wind.svu`

Problem Statement

A popular fast Fourier transform (FFT) application is in performing spectral analysis of a continuous-time signal. In this example we consider an analog deterministic signal composed of three sinusoids. Specifically, the input signal model is of the form

$$s_a(t) = 10\cos[2\pi(1000)t] + 10\cos[2\pi(1100)t] + 0.001\cos[2\pi(3000)t] \quad (1)$$

Note that the first two sinusoids are at 1 kHz and 1.1 kHz, respectively, while the third sinusoid is at 3 kHz and the amplitude is attenuated by 80 dB with respect to the first two sinusoids. Assuming an infinite observation interval, the Fourier transform (in-the-limit) of $s_a(t)$ is

$$S_a(f) = 5[\delta(f - 1000) + \delta(f + 1000)] + 5[\delta(f - 1100) + \delta(f + 1100)]$$
$$+ 0.0005[\delta(f - 3000) + \delta(f + 3000)] \quad (2)$$

The block diagram of a basic Fourier analysis processor is the following[1]:

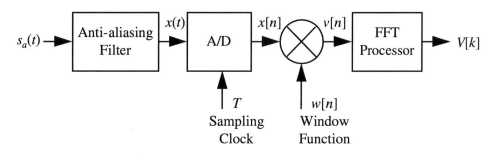

_____

[1] A. Oppenheim and R. Schafer, *Discrete-Time Signal Processing*, Prentice-Hall, Englewood Cliffs, NJ, 1989, p. 696.

SystemView Student Edition

Figure 1: Basic FFT-based spectrum analyzer.

The antialiasing filter is used to remove (minimize) aliasing that results from uniform sampling of the analog signal $s_a(t)$. The window function (sequence) converts what is likely to be a very long-duration sequence into a finite-length sequence that can be processed by the FFT. For this example, the sampling rate is chosen to be 10 kHz, thus aliasing is not a problem and the antialiasing filter can be omitted. The input to the FFT processor is of the form

$$v[n] = x[n]w[n] = s_a(nT)w[n] \qquad\qquad (3)$$

The Fourier transform of sequence $v[n]$ is

$$V(e^{j\omega}) = \frac{1}{2\pi} \int_{-\pi}^{\pi} X(e^{j\theta})W(e^{j(\omega-\theta)})d\theta \qquad\qquad (4)$$

where $W(e^{j\omega})$ is the Fourier transform of the window function. Assuming an N-point, FFT is performed:

$$V[k] = V(e^{j\omega})\Big|_{\omega = 2\pi k/N}, \quad 0 \le k \le N-1 \qquad\qquad (5)$$

Note that each FFT *bin* frequency corresponds to analog frequency variable f as follows:

$$f_k = \frac{k}{NT} = \frac{kf_s}{N}, \quad 0 \le k \le N-1 \qquad\qquad (6)$$

where $f_s = 1/T$ is the sampling rate.

As a practical matter, only a finite number of signal samples can be collected for use by the FFT processor. The spectrum $V[e^{j\omega}]$ is thus a distorted version of the true spectrum due to the spectral spreading or *leakage* that results form the periodic convolution operation of (4). For a sum of unaliased sinusoids such as in (1), the theoretical spectrum is

$$V(e^{j\omega}) = \sum_{i=1}^{3} \frac{A_i}{2} \left[W(e^{j(\omega - \omega_i)}) + W(e^{j(\omega + \omega_i)}) \right] \qquad (7)$$

where

$$A_1 = 10, \omega_1 = 2\pi(0.1), \ A_2 = 10, \omega_2 = 2\pi(0.11), \ A_3 = 0.001, \omega_3 = 2\pi(0.3)$$

The default window function is simply a unity weighting of all input samples on the interval $[0, N]$. The Fourier transform of this *rectangular* window has a peak sidelobe level of 13 dB down and a sidelobe rolloff rate of only 6 dB per octave The spectral leakage imposed by this window is severe. A weak sinusoid close to a strong sinusoid will be masked over by sidelobes from the strong signal. The mainlobe does, however, have zeros located just one bin frequency interval away, thus equal amplitude sinusoids can be resolved when very closely spaced. Many nonuniform weight window functions exist that allow the engineer to reduce the spectral leakage with the penalty of decreased spectral resolution (wider mainlobe). One of these window functions is the *Hanning* window

$$w_{Hanning}[n] = \begin{cases} 0.5\{1 - \cos[2\pi n/(N-1)]\}, & 0 \le n \le N-1 \\ 0, & \text{otherwise} \end{cases} \qquad (8)$$

In the frequency domain, the Hanning window features a peak sidelobe level of 32 dB down, a rolloff rate of -18 dB per decade, and a mainlobe width (zero-to-zero) of four frequency bins.

SystemView Simulation

The spectral analysis system described above will now be implemented in SystemView. Very little work is required in creating this simulation since the SystemView Analysis Mode contains the required functionality within the various buttons and pull-down menus. The main task is to define the input signals as given in (1) and then to set the sampling rate and the record length N. The SystemView block diagram is shown in Figure 2.

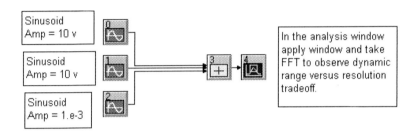

Figure 2: SystemView simulation block diagram.

To start with, the record length is set to $N = 1024$. The simulation is run and then upon switching to the Analysis view, the FFT button is clicked and the vertical display is set to log (FFT magnitude is set by default). The default window mode is rectangular, so the FFT spectral estimate will have the narrow-mainlobe, high-sidelobe properties of the rectangular window. This is verified in the plot of Figure 3 given below.

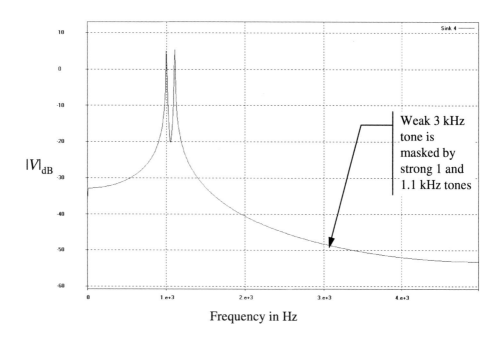

$|V|_{dB}$

Frequency in Hz

Figure 3: Rectangular window spectral estimate
in dB for sinusoids at 1000, 1100, and 3000 Hz, $N = 1024$.

The weak 3-kHz sinusoid is buried in the leakage of the strong 1- and 1.1-kHz tones. In SystemView, we now replot the time-domain waveform and apply a Hanning window from the Window pull-down menu. The FFT is again commuted and the log display mode is invoked. The resulting spectrum is shown in Figure 4.

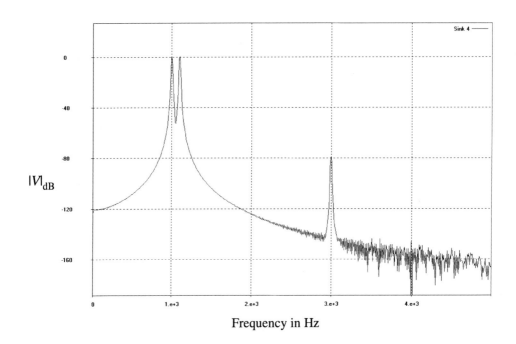

Frequency in Hz

Figure 4: Hanning window spectral estimate
in dB for sinusoids at 1000, 1100, and 3000 Hz, $N = 1024$.

The 3-kHz tone is now clearly evident along with a widening of the mainlobe. To more clearly see the sidelobe structure, *zero padding* of the 1024-point record is required. The SystemView Analysis mode does not provide a direct way of doing this. By adding additional tokens to the block diagram, this feature can be implemented.

The importance of record length is now explored by reducing N to 256. The rectangular window spectral estimate is shown in Figure 5.

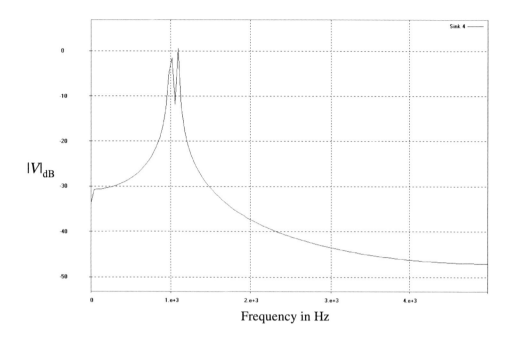

$|V|_{dB}$

Frequency in Hz

Figure 5: Rectangular window spectral estimate
in dB for sinusoids at 1000, 1100, and 3000 Hz, $N = 256$.

By reducing the record length by a factor of four, the resolution is similarly reduced. The closely spaced sinusoids are now more difficult to discern. The corresponding Hanning window spectral estimate is shown in Figure 6.

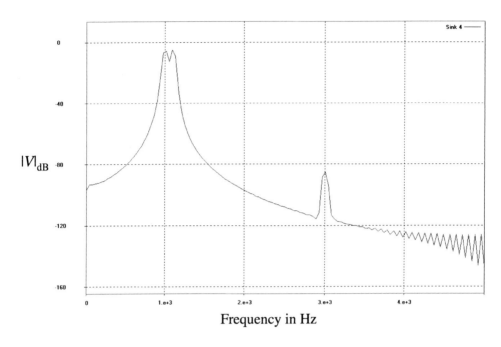

Frequency in Hz

Figure 6: Hanning window spectral estimate
in dB for sinusoids at 1000, 1100, and 3000 Hz, $N = 256$.

The 3 kHz-tone is again visible, but now the increased mainlobe width has nearly merged the two closely spaced tones into one.

Chapter 21. Averaged Periodogram Spectral Estimation

SystemView File: `dsp_aper.svu`

Problem Statement

For a wide-sense stationary random signal, $x(t)$, a frequency-domain representation is obtained from the power spectral density. In the *Wiener-Khinchine* theorem the well-known result that the power spectral density is the Fourier transform of the autocorrelation function is established, i.e.,

$$S_x(f) = F\{R_{xx}(\tau)\} \tag{1}$$

where $R_{xx}(t)$ is the autocorrelation function of $x(t)$.[2] In order to estimate the power spectral density from a single sample function of a random process, it is instructive to consider the power spectrum definition, which states that

$$S_x(f) = \lim_{T_L \to \infty} \frac{E\{|X_{T_L}(f)|^2\}}{T_L} \tag{2}$$

where $X_{T_L}(f)$ is the Fourier transform of a T_L-second segment of $x(t)$. Note that (2) has units of Watts/Hz. An estimator for the power spectrum is the *periodogram*[3]

$$I_{xx}(f) = \frac{1}{T_L}|X_{T_L}(f)|^2 \tag{3}$$

[2] R. Ziemer and W. Tranter, *Principles of Communications*, Fourth Edition, Houghton Mifflin, Boston, MA, 1995, p. 347.
[3] A. Oppenheim and R. Schafer, *Discrete-Time Signal Processing*, Prentice-Hall, Englewood Cliffs, NJ, 1989, p. 731.

which also has units of Watts/Hz. The discrete-time equivalent to $X_{T_L}(f)$ is the L-point discrete Fourier transform (DFT) $V[k]$, $0 \le k \le L-1$ and the associated discrete-time periodogram

$$I_x[k] = \frac{T}{L}|V[k]|^2 , \ 0 \le k \le L-1 \qquad (4)$$

where the index k corresponds to frequency samples kf_s/L, $0 \le k \le L-1$, and f_s is the sampling rate in Hz.

Since the expectation operator is dropped in the definition of the periodogram, the periodogaram of (4) is a sequence of random variables. A single periodogram forms an inconsistent estimate of the true power spectral density. The mean of (4) approaches the true power spectrum as L becomes large, but the variance of (4) remains finite. To reduce the variance of the periodogram spectral, estimate we may average periodograms computed over K, possibly overlapping, data segments. In this example the segments will be contiguous. The averaged periodogram is defined by

$$\bar{I}_{xx}[k] = \frac{1}{K}\sum_{k=0}^{K-1} I_x^k[k] \qquad (5)$$

where the superscript k denotes the kth segment periodogram. A discussion of enhancements to (5), such as window functions and the use of overlapping segments, can be found in S. Kay, *Modern Spectral Estimation* (Prentice-Hall, Englewood Cliffs, NJ, 1988).

SystemView Simulation

In SystemView an averaged periodogram spectral estimator can be implemented with just a few tokens, as shown in Figure 1.

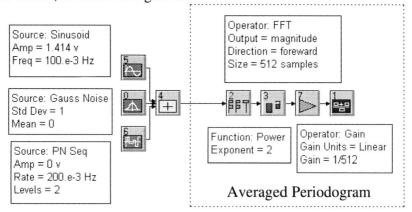

Figure 1: SystemView block diagram.

The key to implementing the algorithm is the averaging sink token (token 7). The System Loop time parameter in the Set System Time dialog box (the clock) determines the parameter K in (5). The averaging sink forms a block-by-block running average over the set of samples computed during each loop. In the analysis window, averaged periodogram results can thus be observed for averages running from one up to K. A disadvantage in placing the FFT token in the simulation block diagram is that a properly scaled frequency axis cannot be obtained. To get as close as possible to a usable frequency scale, the sampling rate is first set to unity and all signal and system parameters are normalized accordingly. Secondly, the simulation start time is backed up onto the negative time axis so that the final $L/2$ points of the kth record (simulation loop) is indexed in the averaging sink from 0 to $L/2$. Note L must be a power of two to conform the radix-2 FFT token record length. The last $L/2$ + 1 samples displayed in the averaging sink correspond to normalized frequency values $f_k = k/L$, $0 \le k \le 1/2$. Assuming that the original input frequency parameters were normalized by a sampling rate of f_s, this [0, 1/2] normalized frequency interval corresponds to the interval $[0, f_s/2]$.

SystemView Student Edition

Two test signals are also incorporated into the simulation of Figure 1:

- A single real sinusoid in additive white Gaussian noise, tokens 0 and 5
- A binary antipodal bit sequence, token 6

The sinusoid plus noise signal model is of the form

$$x[n] = A\cos[2\pi f_0 n + \theta] + w[n] \tag{6}$$

where $f_0 \in (0, 1/2)$, uniform on $[0, 2p)$ random phase q is taken to be zero in the simulation and $w[n]$ is a white Gaussian random sequence with variance s_w^2. The signal-to-noise ratio (SNR) is defined as

$$\text{SNR} = \frac{A^2}{2\sigma_w^2} \tag{7}$$

The theoretical power spectrum of (6) is

$$S_x(f) = \frac{A^2}{4}[\delta(f - f_0) + \delta(f + f_0)] + \sigma_w^2, \quad -1/2 \le f < 1/2 \tag{8}$$

In the simulation results that follow, $f_0 = 0.1$, $A = \sqrt{2}$, and $s_w^2 = 1$, so $\text{SNR}_{\text{dB}} = 0$ dB. The FFT length is $L = 512$ and the number of averages is $K = 10$. The estimated power spectrum of (6) with one average and then ten averages is shown in Figures 2 and 3, respectively.

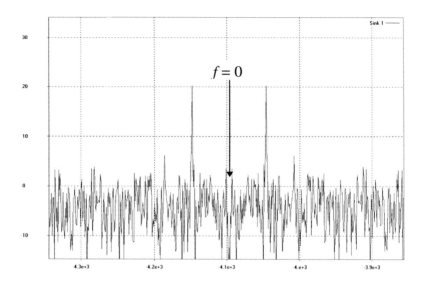

Figure 2: Spectral estimate of one sinusoid in AWGN with $K = 1$, $L = 512$, and SNR = 0 dB.

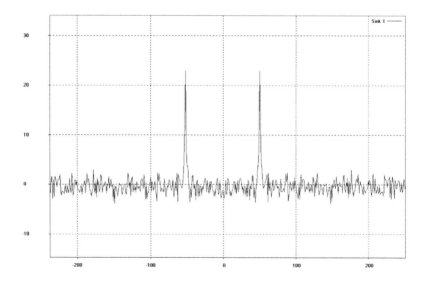

Figure 3: Spectral estimate of one sinusoid in AWGN
with $K = 10$, $L = 512$, and SNR = 10 dB.

The spectral peak due to the sinusoid with amplitude $\sqrt{2}$ should be $10 \log_{10}(512) +$ $\text{SNR}_{\text{dB}} - 3 \text{ dB} = 24.1 \text{ dB}$ above the noise floor. From Figure 3 we see that the simulation result is close to this value. The input SNR of 0 dB is misleading since it does not account for the fact that most all of the power of the sinusoid passes through a single FFT frequency bin and the noise spectrum is uniformly distributed as a power density across all the frequency bins. The variance reduction capability offered by periodogram averaging is clearly visible in the composite plot of all the spectral estimates, $K = 1$ to 10, shown in Figure 4.

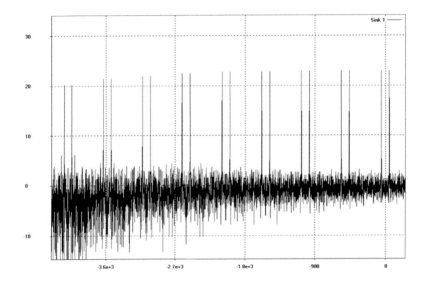

Figure 4: Contiguous spectral estimates of one sinusoid
in AWGN with $K = 1$ to 10, $L = 512$, and SNR = 10 dB.

The second random signal investigated is a time sampled version of the random binary data sequence

$$x_a(t) = A \sum_{k=-\infty}^{\infty} a_k p(t - kT_b - \Delta) \tag{9}$$

where a_k is a sequence of independent, identically distributed random variables equally likely taking on values of ± 1, T_b is the bit duration, $p(t)$ is a rectangular pulse shape function, and Δ is independent of the a_k's and uniformly distributed on the   interval $[-T_b/2, T_b/2]$. In Ziemer and Tranter it is shown that

$$S_{x_a}(f) = A^2 T_b \mathrm{sinc}^2(fT_b) = A^2 T_b \frac{\sin(\pi fT_b)}{\pi fT_b} \qquad (10)$$

When $x(t)$ is sampled by letting $t \to nT$, the power spectrum of (10) is replicated at all multiples of the sampling clock, $f_s = 1/T$, and scaled by $1/f_s$. The result is

$$S_x(f) = \frac{A^2 T_b}{f_s} \sum_{k=-\infty}^{\infty} \mathrm{sinc}\left[(f-k)T_b f_s\right] \qquad (11)$$

where f now denotes the sampling rate normalized frequency.

In the results that follow $A = 1$, $T_b f_s = 0.2$, $L = 512$, and the maximum value of K is 10. The estimated power spectrum of a binary antipodal data sequence for $K = 1$ and 10 is shown in Figures 5 and 6, respectively.

160

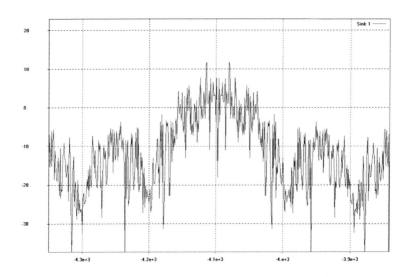

Figure 5: Binary antipodal data sequence power spectrum estimate with $K = 1$ and $L = 512$.

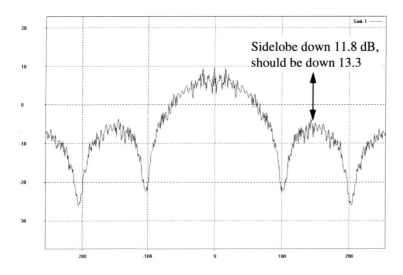

Figure 6: Binary antipodal data sequence power spectrum estimate with $K = 10$ and $L = 512$.

The first positive frequency axis spectral null should occur at about $0.2 \times 512 = 102.4$, which agrees with Figure 6. The sidelobes of the sinc function are slightly higher than the continuous-time result of (10) due aliasing.

Chapter 22. Linear Constant Coefficient Differential Equation Modeling

SystemView File: `con_deq.svu`

Problem Statement

Linear constant coefficient differential equations (LCCDE)s are a popular starting point in the modeling of physical systems. The general LCCDE form considered in this example is

$$\sum_{k=1}^{N} a_k \frac{d^k y(t)}{dt^k} = \sum_{k=1}^{M} b_k \frac{d^k x(t)}{dt^k} \tag{1}$$

The solution of (1) can be accomplished using *classical* techniques as well as the Laplace transform. Simulation of (1) can also provide added insight since the output response can be easily obtained for a variety of input forcing functions and parameter variations. For this example the following third-order equation is studied

$$\frac{d^3 y(t)}{dt^3} + 8\frac{d^2 y(t)}{dt^2} + 17\frac{dy(t)}{dt} + 10y(t) = \frac{dx(t)}{dt} + 3x(t), \quad t \geq 0 \tag{2}$$

Closed-Form Solution

An exact solution to (2) can be obtained using Laplace transform techniques. The system function is

$$\frac{Y(s)}{X(s)} = H(s) = \frac{s+3}{s^3 + 8s^2 + 17s + 10} \tag{3}$$

In control systems the step response is often of interest, so let $x(t) = u(t)$; then we can write

$$Y(s) = \frac{s+3}{s(s^3 + 8s^2 + 17s + 10)} = \frac{s+3}{s(s+1)(s+2)(s+5)} \quad (4)$$

To obtain $y(t)$, expand (4) using partial fractions

$$Y(s) = \frac{K_1}{s} + \frac{K_2}{s+1} + \frac{K_3}{s+2} + \frac{K_4}{s+5} \quad (5)$$

where it is easily shown that $K_1 = 3/10$, $K_2 = -1/2$, $K_3 = 1/6$, and $K_4 = 1/30$. The step response is thus

$$y(t) = \frac{3}{2}u(t) - \frac{1}{2}e^{-t}u(t) + \frac{1}{6}e^{-2t}u(t) + \frac{1}{30}e^{-5t}u(t) \quad (6)$$

Simulation Model

To simulate (2) in an *analog computer*-like fashion, we begin by integrating both sides of (2) three times and isolating $y(t)$:

$$y(t) = -8\int y(t)dt - 17\int\int y(t)dt - 10\int\int\int y(t)dt$$
$$+ \int\int x(t)dt + 4\int\int\int x(t)dt \quad (7)$$

The integral equation form of (7) has the block diagram representation given in Figure 1.

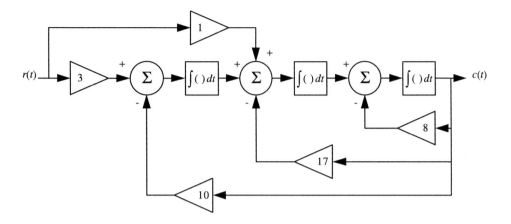

Figure 1: Block diagram representation of the third-order LCCDE.

SystemView Simulation

To simulate (2) in SystemView, the approach of Figure 1 can be implemented or we may directly represent the s-domain system function using a Laplace linear system token. In this example both approaches are taken so as to verify the theoretical equality. In the simulation some small numerical differences may exist. As an additional analysis check, the closed-form solution of (6) is also represented in SystemView. The SystemView block diagram containing three separate simulation systems is shown in Figure 2.

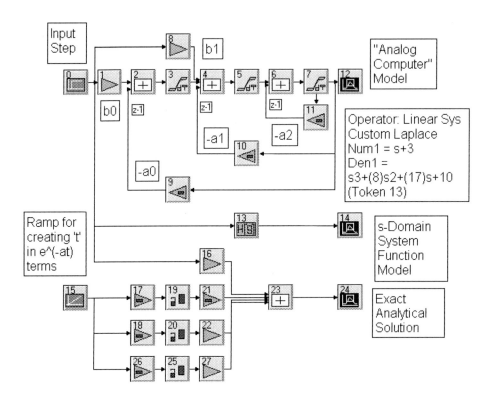

Figure 2: SystemView simulation block diagram.

The system function given by (3) is loaded into token 13 as a ratio of polynomials in *s*. The Laplace System Design dialog box from the linear system token is shown in Figure 3.

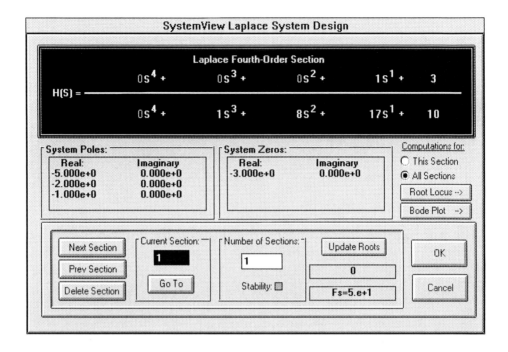

Figure 3: Laplace System Design dialog box in which the system function of (3) is specified.

Note the poles and zeros are found automatically by SystemView and agree with the factoring in the right side of (3).

The choice of sampling rate influences the accuracy of the simulated time-domain response. SystemView performs a digital simulation of an analog system by approximating the differential equation with a difference equation. As evidenced by Figure 3, there is more than one way to do this. An obvious concern is numerical errors. In Phillips and Harbor it is pointed out that numerical solution errors result from

1. The difference equation solutions only approximating the differential equation solutions

2. Computer roundoff errors. [1]

---

[1] C. Phillips and R. Harbor, *Feedback Control Systems*, Second Edition, Prentice-Hall, Englewood Cliffs, 1991, p. 100.

SystemView Student Edition

In SystemView, the integrator token represents analog integration as either *zero-order*, which is rectangular integration, and *first-order*, which is trapezoidal integration. In this example, the zero-order option is used. SystemView converts custom Laplace system functions to difference equation form using the *bilinear transformation*. As a point of interest, the bilinear transformation of an analog integrator is equivalent to trapezoidal integration. Increasing the sampling rate, or, equivalently, decreasing the time step parameter in the numerical solution usually allows the numerical algorithm to better approximate the true solution. Phillips and Harbor point out that when the time step becomes too small, roundoff errors may actually result in increased simulation error.

The simulation block diagram of Figure 1 allows the comparison of two different numerical solution techniques with the theoretical solution. In this example, the smallest time constant is 1 s. The time step is chosen to be 0.02 s. The step system response as obtained from the two simulation models and the exact response are shown in Figure 4.

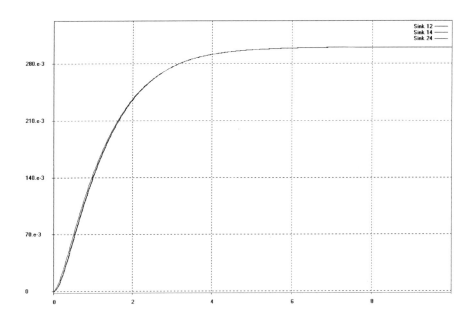

Figure 4: SystemView-generated step response plots for a time step of 0.02 s.

At this resolution all three curves appear to lie on top of each other. In Figure 5 the rising edge of the step response is zoomed.

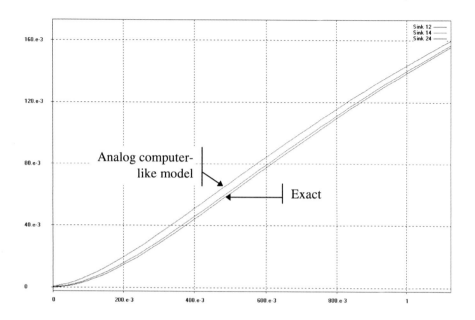

Figure 5: Zoomed step response showing the exact results on the bottom, the H(s) model in the middle, and the analog computer-like model on the top.

In Figure 5 we see that the *H(s)* model, which you recall uses the bilinear transformation, is very close to the exact step response. By decreasing the time step to 0.01 s, the error in the analog computer-like model is reduced, but is still larger than the *H(s)* model.

Parameter Stepping

For multiple loop simulations, SystemView has the capability of stepping parameter values in certain types of tokens. Furthermore, at the start of each simulation loop the initial system conditions may be reset to zero. As an example of parameter stepping, we will step the second integrator feedback gain coefficient over the values –17, –14, –10, –8, and –6 with a five-loop simulation. By having SystemView reset initial conditions back to zero at the end of each loop, the step response is again simulated during each loop. Sink token 12 holds the concatenation of five step responses, each with a different feedback parameter. To overlay the step responses on top of each other, we use the time slice option available in the Analysis window. The parameter step results are shown in Figure 6.

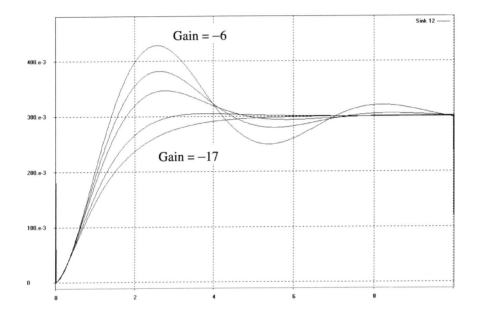

Figure 7: Parameter step results for the analog computer-like system when the second integrator feedback gain takes on values of −17, −14, −10, −8, and −6.

Analysis of the system poles would reveal that as the coefficient gain decreases from −17 to −6, a pair of real poles split off from the negative real axis of the s-plane and move toward the $j\omega$-axis.

Further Investigations

At this point many possibilities exist for further investigation. One area is to verify the simulation accuracy of other easily solvable linear time-invariant systems in a test bed similar to that used here. As a starting point, consider the second-order system function

$$\frac{Y(s)}{X(s)} = \frac{\omega_n^2}{s^2 + 2\zeta\omega_n s + \omega_n^2} \tag{8}$$

where ω_n is the natural frequency and ζ is the damping factor. The step response can be shown to be

$$y_s(t) = \left[\frac{1}{\omega_n^2} - \frac{1}{\omega_n^2\sqrt{1-\zeta^2}} e^{-\zeta\omega_n t} \sin\left(\omega_n\sqrt{1-\zeta^2} + \cos^{-1}\zeta\right) \right] u(t) \tag{9}$$

For convenience, choose $\omega_n = 1$ and for various values of $\zeta < 1$ and simulation step times compare the accuracy of the two simulation implementations compared to (9).

In control system work, the peak value of the step response and the time at which it occurs are of interest. The peak value in theory occurs at time

$$T_p = \frac{\pi}{\omega_n\sqrt{1-\zeta^2}} \tag{10}$$

and the percent overshoot is

$$P_o = e^{-\zeta\pi/\sqrt{1-\zeta^2}} \times 100\% \tag{11}$$

Experimental verification of these formulas can be carried out using the parameter step features of SystemView.

Chapter 23. Control System Design using the Root-Locus

SystemView File: `con_rloc.svu`

Problem Statement

In this example root-locus methods are used to check stability and determine the gain setting in a feedback control system. The system block diagram is shown in Figure 1.

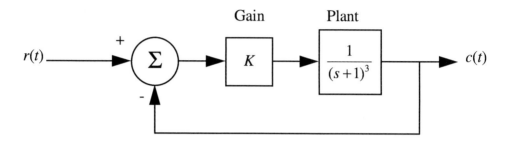

Figure 1: Control system block diagram with free variable K.

The first concern is stability. The values of K that yield a stable system can be found by constructing the Routh array.[2] The system of Figure 1 has a closed-loop system function of the form

$$\frac{C(s)}{R(s)} = H(s) = \frac{\dfrac{K}{(s+1)^3}}{1+\dfrac{K}{(s+1)^3}} = \frac{K}{(s+1)^3 + K} = \frac{K}{s^3 + 3s^2 + 3s + 1 + K} \quad (1)$$

---

[2] C. Phillips and R. Harbor, *Feedback Control Systems*, Second Edition, Prentice-Hall, Englewood Cliffs, NJ, 1991, p. 187.

The Routh array for the polynomial denominator is

$$
\begin{array}{c|ccc}
s^3 & 1 & 3 & 0 \\
s^2 & 3 & +K & 0 \\
s & -K)/3 & 0 & 0 \\
1 & +K & 0 & 0
\end{array}
$$

For stability we must have

$$
\left.\begin{array}{l}
\dfrac{8-K}{3} > 0 \rightarrow K < 8 \\[2ex]
1+K > \rightarrow K > -1
\end{array}\right\} \Rightarrow -1 < K < 8 \qquad (2)
$$

SystemView Simulation

The SystemView simulation block diagram is shown in Figure 2.

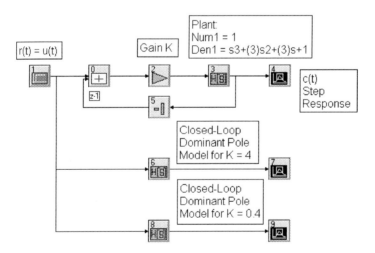

Figure 2: SystemView block diagram of the third-order system along with dominant pole approximations for two values of K.

In Figure 2 the third-order closed system of Figure 1 is implemented directly using a Laplace system token. Below the feedback system are two additional Laplace system tokens that implement second-order dominant pole approximations to the third-order system. These two systems will be discussed in more detail below.

The root-locus of the system is found in SystemView by opening the loop at the feedback summer junction. The loop gain, K, is also set to unity so that the root-locus plot will track the system K value correctly. Following the selection of the s-domain root-locus option, SystemView will ask the user to select the token corresponding to the last token in the open-loop system. The inverting token is chosen in this case. The resulting root-locus plot is shown in Figure 3.

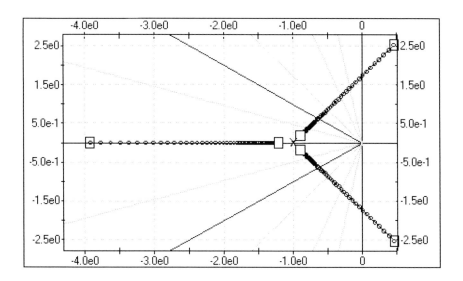

Figure 3: Root-locus plot for log increments in K from 10^{-3} to 25.

By moving the mouse cursor over the root-locus and reading the gain display, it is verified that the complex pole pair does indeed cross the $j\omega$-axis at $K = 8$. The Bode plot of the open-loop system, including the phase margin calculation, is shown in Figure 4.

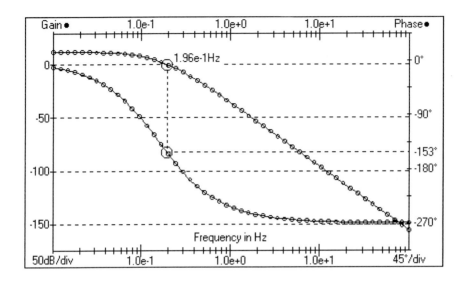

Figure 4: Bode plot of the open-loop system showing the phase margin.

To create this plot, SystemView again asks for the last token in the open-loop system. In this case the Laplace token is chosen.

Dominant Pole Approximation

Recall that the three open-loop poles are located at $s = -1$. As K increases, there is a real pole that moves down the negative real axis. In particular, when $K = 4$ the conjugate pole pair is located at $p_{1,2} = -0.2063 \pm j1.3747$ and the real pole is located at $p_3 = -2.5874$. An analysis task might be to determine the step response overshoot and the time that the peak overshoot occurs. With the SystemView simulation constructed this is a simple matter. Before running the simulation, consider a dominant pole approximation to reduce the third-order system to a simple second-order system. In DiStefano et al. the second-order approximation is shown to be reasonable provided

$$\left| p_r \right| > 5 \left| \operatorname{Re} p_c \right| \quad \text{for } \zeta > 0.5 \tag{3}$$

where p_r is the real pole and p_c is one the complex conjugate poles, and ζ is the second-order term damping, which is the cosine of the angle the poles make to the negative real axis.[3] Here the angle is $81.46°$ and $\zeta = 0.148$, so the approximation is not valid without error.

We now proceed to form the second-order approximation and will use SystemView to check the error involved in the approximation. The system function, for $K = 4$, is approximated as follows:

$$H(s) = \frac{4}{s^3 + 3s^2 + 3s + 5} = \frac{4}{(s + 2.5874)(s^2 + 0.4126s + 1.9324)}$$

$$\approx \frac{4/5 \cdot 1.9324}{s^2 + 0.4126s + 1.9324} = \frac{1.5459}{s^2 + 0.4126s + 1.9324} \tag{4}$$

Note that gain scaling is required to ensure the same dc system gain. Simulation results comparing the step response of the actual third-order system with the dominant second-order approximation are shown in Figure 4.

---

[3] J. DiStefano et al., *Feedback and Control Systems*, Second Edition, Schaum's Outline Series, McGraw-Hill, NewYork, NY, 1990, p. 348.

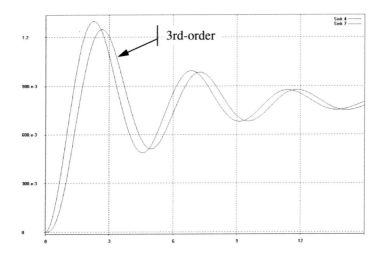

Figure 4: Step response with $K = 4$ for the third-order system and dominant second-order approximation.

The second-order system approximation has peak overshoot given by

$$M_o = A\left(1 + e^{-\pi\zeta/\sqrt{1-\zeta^2}}\right) \tag{5}$$

where A is the step response final value–here, $A = 4/5$. The time where this peak occurs is given by

$$T_p = \frac{\pi}{\omega_n\sqrt{1-\zeta^2}} \tag{6}$$

In the second-order approximation, $\zeta = 0.1484$ and $\omega_n = 1.3901$. The overshoot and peak time values obtained from the simulation and (5) and (6) are compared in Table 1 below.

Table 1: Peak overshoot and peak time comparisons for $K = 4$.

	Second-Order Theory	Second-Order Simulation	Third-Order Simulation
M_o	1.299	1.299	1.272
T_p	2.29 s	2.25 s	2.65 s

The results above indicate that the second-order approximation has more overshoot and a shorter time is required to reach the peak. This is expected from the analysis presented in DiStefano.

As a more design-oriented task, consider finding a gain value K that gives an equivalent second-order damping factor of 0.707. From the root-locus plot of Figure 2 we need to find the root-locus intersection with the 45° constant damping line in Figure 2. Using the mouse and observing the gain display in the SystemView root-locus display, we find that $K = 0.4$ is appropriate. In a practical sense, a compensator would likely be introduced before using a loop gain this small. Continuing with the design, setting $K = 0.4$ places the complex poles at $p_{1,2} = -0.6308 \pm j0.6419$ and $p_3 = -1.739$. Although the dominant pole damping condition of (3) is now satisfied, $1.739/0.6308 = 2.7568 < 5$, so the real pole will have significant influence on the step response.

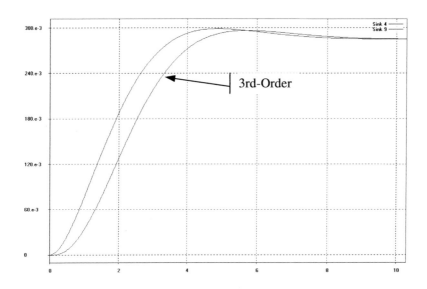

Figure 5: Step response with $K = 0.4$ for the third-order system
and dominant second-order approximation.

In the second-order approximation, complex pole locations imply that $\zeta = 0.701$ and $\omega_n = 0.9$. The overshoot and peak time values obtained from the simulation and (5) and (6) are compared in Table 2 below.

Table 2: Peak overshoot and peak time comparisons for $K = 0.4$.

	Second-Order Theory	Second-Order Simulation	Third-Order Simulation
M_o	0.299	0.299	0.296
T_p	4.89 s	4.88 s	5.72 s

Surprisingly, the peak overshoot of the third-order system is very close to that predicted by the dominant second-order model. The third-order system, however, takes about 0.8 s longer to achieve the peak value that the dominant second-order model predicts.

Section III. Appendices

Appendix A. SystemView Action Buttons

The icon buttons on the Toolbar in the Design window are the means for interfacing with and managing the tokens in your system. The following list summarizes the function of each button.

Delete

> Used to delete a token or a group of tokens from the Design window. To delete a single token, click Delete with the left mouse button. A square symbol, the action pointer, appears in place of the normal mouse arrow. Using the mouse, place this pointer over the token to be deleted and click once with the left mouse button. The token will be removed from the system. **Tip:** *depress the Ctrl key and click and drag the mouse to delete groups of tokens <u>within</u> the box. If you click and drag with the <u>right</u> mouse button, all tokens <u>outside</u> the box will be deleted.*

Clear System

> Used to clear a system from the Design window. To clear a system, select Clear with the left mouse button. If you have not saved the system, a message will appear asking whether you wish to save. If desired answer yes and the Save menu will appear. You may also Cancel the Clear action. Otherwise the Design window will be cleared.

Cancel

> The button is used to nullify an Action button selection. Cancel is activated by placing the Action pointer, generated by selecting another action button, over the Cancel button and depressing the left mouse button.

Disconnect

> Disconnects two connected tokens. To disconnect, select the Disconnect button with the mouse pointer. After depressing the left mouse button, the Action pointer will appear. Place this pointer over the first of the tokens which you wish to disconnect and depress the left mouse button once. Next place the Action pointer over the second token and again depress the left mouse button. The line connecting the tokens will disappear. TIP: 1) press the Shift key and click a sequence of tokens to bypass the Disconnect button.

2) press the Ctrl key and click and drag the mouse to disconnect groups of tokens.

Connect

Used to connect two tokens. To connect, select the Connect button with the mouse pointer. After depressing the left mouse button, the Action pointer will appear. Place this pointer over the first of the tokens which you wish to connect and depress the left mouse button once. Next place the Action pointer over the second token and again depress the left mouse button. A line will appear connecting the two tokens with an arrow pointing from the first selected to the second. TIP: press the Ctrl key and click a sequence of tokens to bypass the Connect button.

New Note Pad

Inserts a blank Note Pad at the center of the screen. You can now make entries, relocate, and realign the Note Pad in accordance with discussions in Chapter 10.

Params -> Note Pad

Automatically creates a new note pad with the parameters of the selected token placed in the Note Pad. The Note Pad will automatically appear. You don't have to open a new Note Pad first.

Execute

Causes the simulation to be executed. To execute a simulation, place the mouse pointer over this button and depress the left mouse button. If the simulation is properly constructed, the system will execute. Otherwise diagnostic messages will appear prompting you to make changes in your simulation.

View MetaSys

Allows you to view and edit the internal structure of a MetaSystem imbedded in your simulation. To view the MetaSystem, click the View MetaSys button with your mouse. The Action Pointer will appear. Place the pointer over the MetaSystem of interest and depress the left button once. A new Design window will appear showing the MetaSystem. *The MetaSystem token is not functional in the Student Edition. For information on upgrading to the Professional Edition, refer to the information request form at the back of this book.*

Create MetaSys

Create a MetaSystem from a selected group of tokens. Press the Ctrl key and click the left mouse button and drag to create a MetaSystem from the group of tokens within the box. *The MetaSystem token is not functional in the Student Edition. For information on upgrading to the Professional Edition, refer to the information request form at the back of this book.*

Root Locus

Compute the system root locus (see Chapter 15).

Bode Plot

Compute the system Bode plot (see Chapter 15).

Re Draw

Causes the entire System window to be re-drawn. To select Redraw, place the mouse pointer over the Redraw button and depress the left mouse button. The System window will be re-drawn.

Reverse

Reverses the input / output direction of a token. To reverse a connection click the Reverse button. The Action pointer will appear. Move this pointer to the token which requires reversal and click. The connections into and out of the token will be reversed as described above. This action is typically used to make your systems easier to read.

Appendix B. SystemView Menus

<u>File</u>

❑ **New System** Clears the current system.

❑ **Open System** Opens an existing SystemView file for analysis or modification.

❑ **Save System** Saves the contents of the current system using the existing file name. If a file name has not been selected, you will be prompted to enter the desired name in the dialog box.

❑ **Save System As** Saves a system by a name you choose. Use with either a new untitled system or, if you wish, to save a titled system under another name.

❑ **Print System (Text Tokens)** Prints the contents of the system screen. The tokens are replaced with boxes containing textual information on each token.

❑ **Print System (Symbolic Tokens)** Prints the contents of the system screen as you see it.

❑ **Print System Summary** Prints a listing of the system tokens with their parameter values. The Time window data is also printed along with the file name and the date.

❑ **Print Connection List** Prints a listing of the system tokens with their input and output connections identified.

❑ **Print Real Time Sink** After selecting this option, click on any Real Time Sink in your system.

❑ **Print SystemView Sink** After selecting this option, click on any SystemView Sink in your system.

❑ **Printer Setup** Opens a dialog which allows you to specify and configure your printer.

❑ **Printer Fonts**. Opens a dialog which allows you to change the font characteristics of the printer and Note Pads.

❑ **Exit SystemView** Closes SystemView and returns you to Windows.

Edit

❏ **Copy Note Pad** Copies the contents of the selected note pad to the clipboard.

❏ **Copy SystemView Sink** Copies the graphic associated with the selected SystemView Sink to the clipboard.

❏ **Copy System** Copies the current system to the clipboard.

❏ **Copy System: Selected Area** Press the Ctrl key and drag the mouse to copy a selected area of your system to the clipboard.

❏ **Copy System: Text Tokens** Copies the current system to the clipboard as boxes with text labels rather than symbolic tokens.

❏ **Copy Entire Screen** Copies the entire SystemView screen to the clipboard.

❏ **Paste To Note Pad** Pastes the textual contents of the clipboard to the selected note pad.

❏ **Paste to System** Pastes the graphic contents of the clipboard to the system screen.

❏ **Undo System Paste** Removes the selected item which has been pasted to the system.

❏ **Delete** Removes the selected item (token or Note Pad) from the system.

Preferences

❏ **Snap To Grid** Registers tokens to the grid lines on the screen. Helpful in aligning your tokens.

❏ **Finer Snap** Registers tokens to the nearest half square of the grid.

❏ **Show Grid** Displays the grid on the screen as desired.

❏ **Color Coded Connections** Attaches the color of the "from" token to the connection line. This feature is highly recommended for ease of visualizing your token connections.

❏ **Allow Warning Messages** When checked, all SystemView warning messages are enabled (e.g., unconnected tokens).

❏ **Show Implicit Delays** Displays the z^{-1} symbol in your system at the physical location of each implicit delay (see Chapter 13).

❏ **Always Check Connections** When checked, SystemView will check your token connections prior to every system execution.

❏ **Use Color Printer** When checked, all printing and pasting will be in color.

❏ **Set Grid Colors** Opens a dialog which allows you to customize the color of your grid.

❏ **Set Background Colors** Opens a dialog which allows you to customize the background color of your screen.

❏ **Use Default Colors** Return to SystemView default colors from current customized colors.

View

❏ **Zoom** This feature allows you to expand or contract the view of the displayed system. The sub-menu controls have the following meaning:

125%	125% of Normal view.
Normal	
80%	80% of Normal view.
70%	70% of Normal view.
60%	60% of Normal view.
50%	50% of Normal view.

❏ **MetaSystem** Allows you to view the contents of the selected MetaSystem. *The MetaSystem token is not functional in the Student Edition. For information on upgrading to the Professional Edition, refer to the information request form at the back of this book.*

❏ **Hide Token Numbers** Hides the token numbers on the System tokens.

❏ **Calculator** Launches the Windows calculator.

Note Pads

❏ **Hide Note Pads** Hides your note pads from view. They are not deleted, and may be recovered by disengaging this option.

❏ **New Note Pad** Inserts a blank Note Pad at the center of the screen. You can make entries, relocate, and realign the Note Pad in accordance with discussions in Chapter 10.

❏ **Copy Token Parameters to Note Pad** Allows you to generate a note pad with the parameters of the selected token automatically placed in the Note Pad. The Note Pad will automatically appear. You don't have to open a new Note Pad first.

Connections

❑ **Disconnect All Tokens** Selecting this feature will disconnect all tokens in your system. It will not change token parameters.

❑ **Check Connections Now** Clicking on this option performs a one time check of all your system connections.

❑ **Animate Exe Sequence** Clicking on this option causes SystemView to step through the token execution sequence. At each step the currently executing token is highlighted. This feature is useful in conjunction with the execution sequence editor (Chapter 16).

❑ **Use Default Exe Sequence** Clicking on this option causes SystemView to use the execution sequence as determined by the SystemView compiler, rather than your custom execution sequence. This feature is used in conjunction with the execution sequence editor (Chapter 16).

❑ **Use Custom Exe Sequence** Clicking on this option causes SystemView to use your custom execution sequence, rather than the execution sequence as determined by the SystemView compiler. This feature is used in conjunction with the execution sequence editor (Chapter 16).

❑ **Edit Exe Sequence** Clicking on this option gives you a choice of Use Sys Tokens or Use Exe List to edit the system execution sequence (see Chapter 16 for detailed instructions).

❑ **Cancel Edit Operation** Clicking on this option cancels the current execution sequence edit operation (see Chapter 16 for detailed instructions).

❑ **Cancel Last Edit** Clicking on this option cancels only the last execution sequence edit operation (see Chapter 16 for detailed instructions).

❑ **End Edit** Clicking on this option ends the execution sequence edit operation (see Chapter 16 for detailed instructions).

System

❑ **Compile System Now** Selecting this option will force SystemView to re-compile your system without actually executing the simulation.

❑ **Single Step** When selected, the system execution will proceed one step at a time as you tap the keyboard Spacebar. All SystemView tokens, including Sinks, update in "real time" at each step. For example, the Data List Sink will display its new value at each step. This feature is especially useful when debugging your system..

❑ **Debug (User Code)** This menu entry is enabled only if you have purchased the User Code Option. When selected, the Data List and Current Value Sinks will display special debug information as received from each User Code token. *The User Code option is not available for the Student Edition. For information on upgrading to the Professional Edition, refer to the information request form at the back of this book.*

❑ **Root Locus** This menu entry is used to invoke the root locus computations and display for the current system. A shortcut is to click the Root Locus button located on the Action Bar (see Chapter 15).

❑ **Bode Plot** This menu entry is used to invoke the Bode plot computations and display for the current system. A shortcut is to click the Bode Plot button located on the Action Bar (see Chapter 15).

Tokens

❏ **Find Token** Selecting this option will produce a list of all the system tokens by number, along with a brief description of each. Select the token of interest. A red square will flash where the selected token resides in the system.

❏ **Find Implicit Delay Tokens** You can determine where the SystemView compiler enforces delays in a feedback system by selecting this option. When you make this selection a panel will appear listing the token pairs between which a one sample delay exists. To view any of these tokens, select it from the list and click on the button "Go To ...". This action will cause the selected token to flash on the design screen. See Connections: Edit Execution Sequence above, and Chapter 13 of the User's Guide for further information.

❏ **Move Selected Tokens** After selecting this option click and drag the mouse to outline the selected system tokens. The selected tokens will move accordingly. **Tip:** *depress the Ctrl key and drag the mouse to move groups of tokens.*

❏ **Move All Tokens** After selecting this option click and drag on any system token. The entire system will move accordingly. **Tip:** *press the right mouse button while over any token to move all tokens.*

❏ **Duplicate Tokens** After selecting this option click on a token. This token will be duplicated exactly, with all parameter values and placed one half square from the selected token. **Tip:** *depress the Ctrl key and drag the mouse to duplicate groups of tokens.*

❏ **Create MetaSystem** Creation of a MetaSystem is greatly simplified through the use of this feature. The mouse pointer will change to the black rectangle. While pressing the Ctrl key, drag the mouse, to outline the tokens to be converted into a MetaSystem. When you release the mouse button, these tokens will collapse into a single MetaSystem, with MetaSystem input and output tokens automatically added. *The MetaSystem token is not functional in the Student Edition. For information on upgrading to the Professional Edition, refer to the information request form at the back of this book.*

❏ **Rename MetaSystem** Choosing this selection from the pull down menu allows you to rename a MetaSystem within your simulation. When you make this selection, the rectangular cursor appears. Place this over the MetaSystem and click. A panel appears with the current name of the MetaSystem that you may change to suit your needs. *The MetaSystem token is not functional in the Student*

Edition. For information on upgrading to the Professional Edition, refer to the information request form at the back of this book.

❑ **Explode MetaSystem** A MetaSystem can be exploded from a single token to its constituent tokens. The mouse cursor will change to the black rectangle. Simply click the desired MetaSystem. When using this feature, it is advisable to leave sufficient space to the right and below the MetaSystem for expansion. After you create a MetaSystem you can name it using the 'Rename MetaSystem' selection under the Tokens menu. *The MetaSystem token is not functional in the Student Edition. For information on upgrading to the Professional Edition, refer to the information request form at the back of this book.*

❑ **Assign Custom Token Picture** You can assign a custom picture to a SystemView token in the following way. Use your favorite draw program to create the token picture to be assigned. The format may be a bit map (.bmp), an icon (.ico) or a windows metafile, (.wmf). Be sure to size the picture to approximately the same size of the tokens as they appear on the SystemView screen.

❑ **Use Default Token Picture** You can return to the SystemView default token picture by selecting this option. The black rectangular cursor will appear. Place the cursor over the selected token and click.

Help

❑ **Enable Library** This menu entry provides the means for you and ELANIX to enable the optional SystemView libraries (User Code, Communications, etc.) over the telephone. Each optional library can be activated by phone after the appropriate license has been purchased. To enable a library, select Enable Library from the Help pull-down menu. A panel will appear containing three windows. The first will contain a key. When you call ELANIX, this key will be used to set the authorization number and the verification number which are in the second and third windows, respectively. *Optional libraries are not available for the Student Edition. For information on upgrading to the Professional Edition, refer to the information request form at the back of this book.*

❑ **About SystemView** Provides information about your computer configuration, the SystemView version number, your optional library licenses, and information about ELANIX Inc. (including the SystemView customer support telephone and fax numbers).

Index

F

Feedback paths, time delay in, 41
File extension, 12
File, coefficient file header, 21
Filter design, 22
Filters, 17
FIR, filter design, 22
FM Quadrature Detector, 83
Frequency resolution, 14
Frequency Samples, 18
Frequency, FIR specification, 23
Function Library, 44
Function token, 8

H

Hamming filter, 25
Hamming window, 33
Hanning filter, 25
Hanning window, 33
Header, coefficient files, 21
High pass filter, 22
Hilbert transform, 22

I

IIR, 20
IIR filter design, 26
IIR filter design window, 27
Implicit delay, 42
Importing Coefficients, 21
Installing SystemView, 3

L

Laplace system design window, 29,
    31, 32
Laplace systems, 28
Library, Functions, 44
Library, Operator, 48

Library, Sinks, 55
Library, Sources, 52
Linear Constant Coefficient
    Differential Equation Modeling, 161
Linear Phase filter, 26
Linear system design window, 18
Linear system token, 17
Linear system, Laplace, 28
Low pass filter (FIR), 22
Low pass filter design window, 24
Low pass filters, IIR, 26
Lowpass and Bandpass Sampling
    Theory, 123

M

Manual entry of coefficients, 19
Menu bar, 4
Menus, 185
Message area, 6
MetaSystem I/O token, 7
MetaSystem token, 7
MetaVu, 182
Move Selected Tokens, 191
Multiplier token, 8
Multirate Sampling, 137

N

No. samples, 14
Number of samples, 14
Number of system loops, 15

O

Operator Library, 48
Operator token, 8

$System View$
B Y E L A N I X

YES! Please send me further information on the following ELANIX products!

SystemView by ELANIX, the software presented in this text, is a high level conceptual design and evaluation engine embedded in an intuitive, friendly and completely visual design environment. Running under Microsoft Windows with minimal hardware requirements, SystemView enables engineering students and professionals to perform comprehensive analog, digital, and mixed-mode design and analysis of signal processing, communications, and control applications. For further information on the SystemView Professional Edition, please complete this form and FAX it to ELANIX Marketing at (818) 597-1427.

I am interested in:

❏ SystemView Professional Edition
❏ SystemView Communications
   Library
❏ SystemView DSP Library

❏ SystemView RF/Analog Library
❏ SystemView Logic Library
❏ SystemView User Code Option
❏ All of the above

Name _____

Company: _____

Address: _____

_____

City/State/Zip: _____

Phone: _____

FAX: _____

e-mail: _____

For the fastest response, FAX this form to ELANIX at (818) 597-1427.

ELANIX
I N C O R P O R A T E D

5655 Lindero Canyon Road, Suite 721, Westlake Village, CA 91362
Tel +1 818 597-1414 Fax +1 818 597-1427 BBS 818 597-0306
e-mail systemview@elanix.com URL http://www.elanix.com/elanix/

PWS PUBLISHING COMPANY
LICENSE AGREEMENT